THE ANIMAL FAMILY

Catchpole, Clive

THE ANIMAL FAMILY

Consultant Editor, Dr Philip Whitfield
Illustrated by Richard Orr and Michael Woods
Introduction by Desmond Morris

W· W· NORTON & Co· INC
NEW YORK

Text by Dr Clive K. Catchpole
        Dr Frank H. Hucklebridge
        Dr D. Michael Stoddart

Major illustrations by Richard Orr
Line drawings by Michael Woods

The Animal Family was conceived,
edited and designed by
Marshall Editions Limited
71 Eccleston Square
London SW1V 1PJ
© 1979 Marshall Editions Limited

Editor: Jinny Johnson
Designer: Julia Clappison
Researcher: Janet Wilson

First American Edition 1979
All Rights Reserved

Printed and bound in Spain by Printer
industria gráfica sa Sant Vicenç dels Horts
Barcelona
D.L.B. 23–101–1979

Library of Congress Cataloging
in Publication Data

Catchpole, Clive.
   The animal family.

   Includes index.
   1. Animals, habits and behavior of.
2. Familial behavior in animals.

I. Hucklebridge, Frank H., joint author.
II. Stoddart, David Michael, joint author.
III. Whitfield, Philip. IV. Orr, Richard.
V. Woods, Michael. VI. Title.
QL751.C368   574.5   79-19464

ISBN 0-393-01304-9

CONTENTS

INTRODUCTION

by Desmond Morris
Research Fellow, Wolfson College, Oxford

As the human brain grew steadily larger and more complex, during the course of evolution, it posed a special problem for our ancient ancestors. The human infant needed more and more time to mature. Having such a magnificent learning machine inside its skull gave it a tremendous advantage, but it also placed a heavy burden on its parents. The machine had to be programmed. It had to spend many years acquiring information and storing it in the massive memory banks, if it was to function efficiently. During this long process the developing child required help, and the task of being a human parent became increasingly arduous. Care and protection had to be combined with training and education on a scale far beyond that of any other animal species. The human parent had to become a super-parent.

For hundreds of thousands of years, these super-parents have set up their homes and established their care-giving family units across the face of the globe. The details have varied greatly, but the basic elements have remained the same. Even in the most complicated and sophisticated of our modern human communities—our great towns and cities — the fundamental unit has remained the family home, the safe place inside which we nurture our offspring and prepare them to face adult life. We have the technology to change all this, to abandon the old ways and to replace parental love with baby farms and breeding specialists, just as we have become work-specialists with an intricate division of labour. But we have stubbornly refused to do this. As parents we have retained our primeval ways and still live out our ancient family pattern, like our early ancestors, hundreds of generations before us.

We still indulge in courtship, still fall in love, establish a breeding 'nest', cuddle and fondle our babies, feed and clean our infants, teach and discipline our children, before sending them out to confront the adult world. We spend nearly two decades doing it and, as child follows child, the demands become so heavy that they influence our whole pattern of living and shape the fabric of our human societies.

Even our sexual behaviour has become modified to this end. For us, the mating act has become much more than an act of procreation. It has become a bonding device that helps to cement the relationship between the male and female parent, keeping them together during the long years of shared responsibility. In other primates, the female becomes sexually active only at times of ovulation. In many species she develops a sexual swelling that makes her attractive to the males, but this swelling disappears at other times in her sexual cycle. This means that sexual acts are likely to lead to fertilization. When she cannot be fertilized, she is not interested in sex, nor is she interesting to the males.

In the human species all this has changed. The human female shows no obvious change as the moment of ovulation approaches, and she remains sexually attracted to the male, and sexually attractive to him, even when she is not ovulating. The result is that the majority of human sexual acts cannot lead to fertilization. Only the small minority that happen, accidentally, to coincide with the days of ovulation, can result in the production of an offspring.

Because of the enormous importance of family life in human society, it is natural that we should be interested in the family patterns of other species. We marvel at their colourful courtship displays, their intricate nest-building techniques, their ways of giving birth and feeding their young. This book provides a valuable introduction to the great diversity of ways in which the problems of breeding and rearing are solved by different branches of the animal tree. It shows us the basic alternatives of being a 'quantity breeder' or a 'quality breeder'. The quantity breeder produces huge numbers of offspring and then leaves them to fend for themselves, with the result that only a few—but just enough—survive. The quality breeder produces only a few off-spring, but then lavishes a great deal of energy ensuring that those few flourish and grow.

It also shows us the many more subtle alternatives open to the quality breeders. For some, the offspring are protected by spending an unusually long time inside the mother's body before being thrust out into the external world of risks and dangers. For others, there is an extended phase of protective incubation in a pouch or a nest. For others again, there is a lengthy period of parental control even after they have started to move about and explore the environment, with the alert parent keeping an ever watchful eye on their juvenile exploits.

Nowhere is there such a prolonged period of family care as in the human species, but during the shorter spells of parental devotion of other species there is such a variety and intensity of activity that the subject remains an eternally fascinating one for the naturalist. Indeed, it remains so for anyone who has not become blind to the wonders of the animal kingdom, as the beautifully illustrated pages of this book will testify.

Desmond Morris

All the animals and plants alive on earth today, and the countless millions that have preceded them, owe their existence to the fundamental property that indelibly marks out the animate from the inanimate—self-reproduction. Life in all its forms continues only as a result of the reproductive efforts of previous generations.

At the beginning of life on our planet, some 3 to 4,000 million years ago, some large chemical molecules developed the ability to make copies of themselves and, with it, the means of reproduction. The molecules with the power of replication would have survived at the expense of others without this ability, and swept through the 'primeval soup' from which they emerged, like an epidemic.

Some of the replicator molecules must have developed techniques for breaking up others of their kind to provide themselves with building units for enlargement, and acquired the means of using energy-providing materials from the environment. At some stage early on in their evolution the replicators also started to direct the synthesis of servant molecules. They no longer went naked, but surrounded themselves with membranes—they became cells or 'life units'.

This primitive pattern of organization has been the foundation stone of life, for nearly all today's living creatures are composed of cells, survival machines, containing the replicator molecules which build them. In the language of modern biology replicator molecules are deoxyribonucleic acid (DNA) and they have their own capsule, the nucleus, within the cell. The DNA can not only reproduce itself but contains a code which, when transcribed, determines the structure of the individual which it creates by masterminding the cell's 'worker molecules'. Each set of DNA instructions directing the manufacture of one protein is known as a gene, and in all organisms except viruses, bacteria and some primitive plants, genes are strung beadlike together into filamentous bodies called chromosomes.

The first method of reproduction used by living cells was almost certainly devoid of sex. The replicator molecules simply split in two, as did the contents of each cell, and two identical progeny were formed. Such asexual reproduction persists to this day among single-celled animals, but, although it ensures a rapid rate of increase in numbers, it has some disadvantages. In the single-celled creature *Paramecium*, for example, asexual reproduction by simple duplication and splitting (fission) is the normal means of population growth. But if it continues for too long the individuals gradually become less vigorous, their rate of division slows down and the population eventually dies out, perhaps because of

mistakes in copying of the DNA and the inability of survivors to cope with changes in the environment. Only after a sexual process involving the reshuffling of the genetic material by a union of DNA strands does vigour and the ability to survive return.

Organisms advanced from being single-celled creatures to more complex collections of cells specialized to perform different tasks such as movement and feeding. Included in this specialization was the production of cells designed for reproduction which would carry and express the complete instruction for the rebuilding of a totally new organism. This meant that as long as an individual could survive long enough to produce reproductive cells, then its own death through starvation, injury or disease did not threaten the future survival of its genes. The production of specialized reproductive cells might also involve the union of genetic material from different individuals—that is, sexual reproduction.

The mechanics of sexual reproduction are bound up with the behaviour of the genetic material DNA and the way in which it is arranged within each cell. Most of the cells do not have one but two sets of hereditary instructions. These come in the form of pairs of chromosomes. This pairing is known as the diploid—twofold—state. Its great advantage is that a gene on one chromosome that has become changed by an outside influence or an error in copying may be disguised or dominated by its counterpart on the other chromosome of the pair. This may seem strange since gene-changes, and hence variation, enable populations to adapt to changes in the environment. However, genes do not have this overall perspective in mind. They simply replicate and survive; but sitting in a successful survival machine it is as well to have a cover copy to ensure that they make the machine the same way next time.

In both unicellular and many-celled animals and plants the all important complement of instruction-bearing DNA is paired diploid fashion in two-by-two chromosomes. But because sexual reproduction must involve the union of DNA from different individuals, then the cells carrying the DNA—the gametes or germ cells—must have only a single or haploid set of chromosomes so that the normal state of affairs is resumed when the two combine. This combination or fusion is the process of fertilization.

The sexual distinction between male and female can be made only when the two gametes that combine at fertilization are noticeably different. Female gametes, eggs or ova, are produced in small numbers. They cannot move under their own steam, are large and often contain some sort of food reserve to sustain the new individual during its growth and development. Male gametes, spermatozoa or sperm, are made in massive numbers, are able to move toward the egg, and are tiny.

The process by which haploid eggs and sperm are produced is a special sort of cell division called meiosis. The paired chromosomes in the diploid cell become separated so that a single set of chromosomes forms the genetic material of an egg or sperm. During this division, the chromosome pairs come to lie close together, break and exchange equivalent lengths of DNA. This crossing over and recombination is like the shuffling and dealing of a pack of cards between two players. The pattern on the back of the cards is always the same and represents the species to which the genes belong. Each player will receive an equal number of different cards which distinguish his hand from the other.

In the same way, crossing over and recombination at meiosis results in the production of gametes with new sequences of genes. And because no gamete carries exactly the same genetic information as the parent cell from which it came, every individual created as the result of the fusion of male and female gametes at fertilization is genetically unique. (Identical twins are the exception, but they are formed by a simple doubling—an asexual division—of the genetic material after fertilization has taken place.)

The events of meiosis, which must themselves have arisen by chance alterations in the behavioural repertoire of the original replicator molecules, are the key that has unlocked the treasure chest of evolution. Without it there is little possibility that life would have advanced to the staggering variety of forms that exists today. The clue to its importance is the variation of individuals produced as a result of crossing over and recombination of the genetic material, for in a population of varied forms it is likely that some, at least, will be able to adapt to changes in the environment and thus survive to reproduce the next generation.

The factors which favour the evolution of sex seem to be powerful, but is the whole story understood? Did the long-term survival advantages of variation made possible by sex really operate so strongly in favour of those populations possessing it that all other methods of reproduction were totally outstripped? The problem is that although sex, the habit of nearly all animal species, is a potent force in their evolution and continued existence, it does seem to have some disadvantages.

The process of recombination, for example, is bound to break down a successful combination of genes. To continue the card pack analogy, it is like demanding the re-deal of a hand holding four aces. Equally, the production of males, which do not themselves reproduce, is a disadvantage to the survival of a particular set of genes. It would perhaps be more advantageous and straightforward for the world to be populated solely by females who reproduce simply by manufacturing eggs that can develop without being fertilized. This method of procreation—parthenogenesis—does exist in the animal world, generally in parallel with a sexual method of reproduction. It is an expedient, like the asexual splitting of the *Paramecium*, used to increase numbers rapidly. For most species though, the balance is tipped in favour of the long-term merits of sex.

Since sex has become reproduction's rule rather than its exception, remarkably subtle mechanisms have developed to ensure its success. These mechanisms mean that males and females can recognize each other, that mating occurs at the correct time and that the fusion of egg and sperm and the development of a new individual are guaranteed. Sex is the essence of survival, and life continues only as organisms can reproduce their kind. It is the fascinating features of continual regenesis that are the subject of this book.

THE MECHANICS OF SEX

Each individual created by a male and female is unique — with the exception of identical twins. . . . The capacity for change and improvement is the great advantage of sexual reproduction.

Reproducing without sex

Sex is not always a prerequisite of reproduction. Many animals do without it, while some have never acquired the habit. Yet others are versatile enough to reproduce without sex (asexually) in favourable environmental conditions, but use sexual methods in times of hardship to shuffle the genetic pack and give variability, and to produce resistant 'overwintering' stages that can sit out problems. But this choice is a luxury confined to simple creatures. Man and his fellow backboned animals are entirely dependent on sex for their continued existence.

Asexual reproduction is often a special kind of growth, for it results from exactly the same kind of division of cells (mitosis) which generates normal growth processes. In mitosis the cell's genetic material is doubled, and this doubling is followed by a split in the cell substance or cytoplasm. A new cell or collection of cells then takes up a separate existence. When new individuals are produced by mitosis during asexual reproduction, each new creature is an exact copy of the parent from which it came, and this parent is, at the same time, its identical twin. Only mistakes in copying of the genetic material during duplication, or chance alterations in this material (mutations) will create permanent changes that can be passed on to succeeding, asexually produced generations.

When a single-celled creature simply splits into two equal halves, its asexual reproduction is known as fission. The dividing line may be across the organism's middle or from end to end. In unicellular animals that are parasites, and for which safety in numbers is the keynote of survival, the cell may divide many times within a protective case. The case then bursts to release the progeny. In many-celled animals, such as *Hydra*, offspring are produced as buds from the parent. The parent retains its own identity and the new individual becomes detached eventually.

Sex is a risky business, for fertilization of the female sex cell (egg or ovum) by a male sex cell (sperm) cannot be guaranteed. Some animals can avoid this problem by an asexual by-pass called parthenogenesis in which the ovum develops into an adult without fertilization. Parthenogenesis is common in tiny aquatic creatures called rotifers, and in some crustaceans and insects, including aphids, ants and bees. Some rotifers depend entirely on parthenogenesis. Males are unknown and the eggs all hatch into females. In other rotifers several sorts of egg are made. One sort, produced after a meiotic division and hence with half the normal complement of genetic material, develop into males as long as they are not fertilized, while fertilized eggs result in females.

In the laboratory, even the eggs of backboned animals such as frogs, rabbits and sheep can be induced to develop without mating. In the case of the zebra fish, *Brachydanio*, experimental manipulation can produce adults from unfertilized eggs.

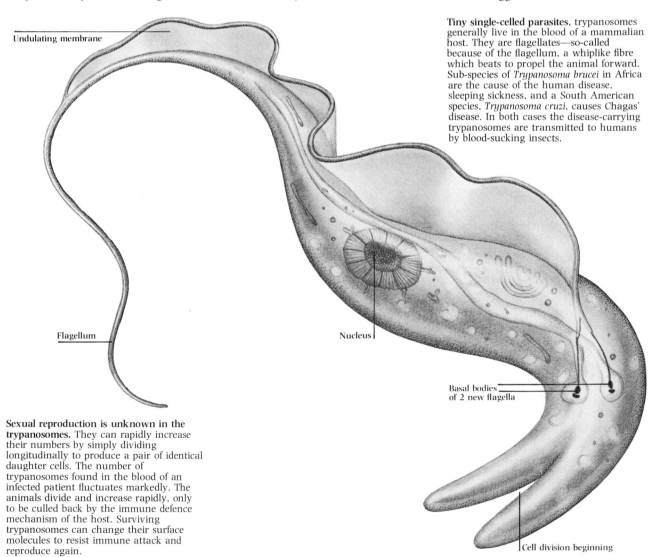

Tiny single-celled parasites, trypanosomes generally live in the blood of a mammalian host. They are flagellates—so-called because of the flagellum, a whiplike fibre which beats to propel the animal forward. Sub-species of *Trypanosoma brucei* in Africa are the cause of the human disease, sleeping sickness, and a South American species, *Trypanosoma cruzi,* causes Chagas' disease. In both cases the disease-carrying trypanosomes are transmitted to humans by blood-sucking insects.

Undulating membrane

Flagellum

Nucleus

Basal bodies of 2 new flagella

Cell division beginning

Sexual reproduction is unknown in the trypanosomes. They can rapidly increase their numbers by simply dividing longitudinally to produce a pair of identical daughter cells. The number of trypanosomes found in the blood of an infected patient fluctuates markedly. The animals divide and increase rapidly, only to be culled back by the immune defence mechanism of the host. Surviving trypanosomes can change their surface molecules to resist immune attack and reproduce again.

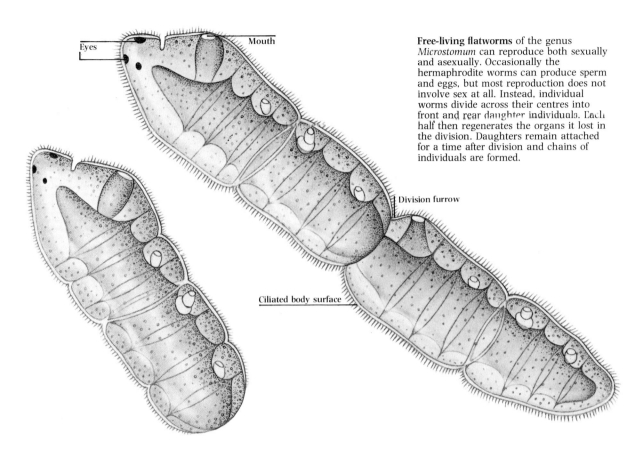

Free-living flatworms of the genus *Microstomum* can reproduce both sexually and asexually. Occasionally the hermaphrodite worms can produce sperm and eggs, but most reproduction does not involve sex at all. Instead, individual worms divide across their centres into front and rear daughter individuals. Each half then regenerates the organs it lost in the division. Daughters remain attached for a time after division and chains of individuals are formed.

Eyes

Mouth

Division furrow

Ciliated body surface

Budding is a form of asexual reproduction. A slight bump appears on the parent's body, and grows to become a miniature hydra. When it is mature it constricts its base and separates from the parent hydra.

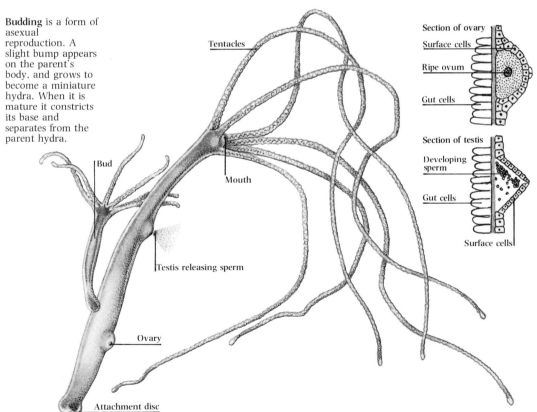

Tentacles

Mouth

Bud

Testis releasing sperm

Ovary

Attachment disc

Section of ovary

Surface cells

Ripe ovum

Gut cells

Section of testis

Developing sperm

Gut cells

Surface cells

Testes and ovaries develop in hydras under certain conditions. Some species have either testes or ovaries but others are hermaphrodite and have both. When mature, the egg cell remains attached to the hydra for a time. Sperm cells are released from the testes and swim to reach and fertilize an egg. Sexual reproduction usually occurs in the autumn. Fertilized eggs are coated in a protective shell and can survive the winter.

Species of the coelenterate *Hydra* are found in fresh water all over the world. The hydra lives attached to vegetation or the stream bed and is usually about 0.6 in (15 mm) long. It has a tubular body with a mouth at the free end surrounded by tentacles. Hydras reproduce sexually or asexually depending on the environment. In the warm summer months hydras reproduce asexually by budding—the new individual grows on the parent's body. Hydras also possess remarkable powers of regeneration. If the body is cut into several pieces, each grows into a complete, but tiny, new hydra.

The units of life

Living matter is packaged in small units or cells, each with the capacity for survival and self-renewal. The director of these processes is deoxyribonucleic acid, DNA, the chemical of heredity. Cell structure, behaviour and biochemistry are ordered by the transcription of messages coded in DNA. Cell size is limited by the area over which DNA can exert its authority, so large organisms comprise millions of cells, each with an exact replica of the DNA of its neighbours. The precise DNA code common to all of a creature's cells is its genotype. Cells may become specialized for particular tasks by utilizing only some of their DNA code, but every cell still retains a complete copy of DNA instructions.

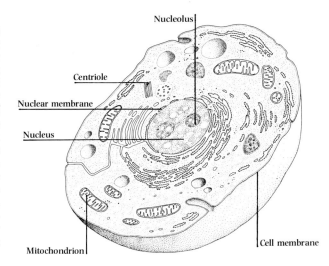

Nucleolus

Centriole

Nuclear membrane

Nucleus

Mitochondrion

Cell membrane

Each cell is a separate entity enclosed within a membrane via which it interacts with its environment and its neighbours. The cell's DNA, whose coded instructions are translated to direct all cell activities, is gathered in a membrane-bounded nucleus. The remaining cell substance, the cytoplasm, is packed with yet more cell components which help organize chemical reactions.

The millions of cells that comprise a complex adult organism begin life as a single cell. It is the process of cell division which effects the transformation from a single cell to many. And the same process makes possible the asexual reproduction of complete organisms. Fundamental to cell multiplication is the replication of DNA containing the coding for cell structure and activity. Division producing 2 cells with exactly the same chromosomal complement of DNA is called mitosis. The DNA duplication takes place when the cell is resting, but only becomes apparent when cell division begins. The chromosomes condense and can be seen to have split lengthwise into 2 chromatids.

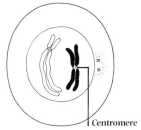

1 Chromosomes appear

2 Chromosomes coil and shorten

Centriole
Spindle

3 Chromosomes attracted to spindle

4 Chromosomes at centre of spindle

In preparation for mitosis the centriole, a structure lying just outside the nucleus, divides. The 2 centrioles migrate to opposite ends (poles) of the nucleus and fibres form between them to make a spindle.

Shortening chromosomes appear as 2 chromatids joined at a constriction, the centromere. The nuclear membrane disintegrates and the chromosomes line up across the spindle centre.

Chromosome behaviour is at the heart of sexual reproduction. In most body cells the chromosomes are found in pairs. In creatures conceived by sexual means, one member of each chromosome pair comes from the mother, the other from the father. The genes in each chromosome pair, although concerned with the same area of instruction, may carry information differing in detail, but because a gene on a chromosome may dominate the equivalent gene on the other chromosome, usually only one version of the genetic order is carried out. For this chromosome pairing—the diploid state—to be ensured, sexual reproduction demands meiosis, a special sort of cell division which results in the production of male and female sex cells or gametes with only a single set of chromosomes. Meiosis thus ensures that when these haploid gametes fuse at the moment of fertilization, the normal diploid state is restored. New, sexually created individuals differ from their parents because meiosis results in new gene combinations, and fertilization in new chromosome pairings.

1 Chromosomes appear

2 Chromosomes associate in pairs

3 Chromatids apparent in chromosomes

Centrioles

Centromere

Spindle

As meiosis begins the chromosomes contract in length and become easier to see.

The 2 members of each chromosome pair move to lie lengthwise alongside one another.

Each chromosome divides into 2 chromatids so that each grouping has 4 chromatids.

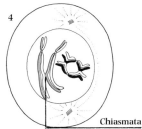

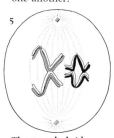

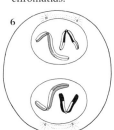

4

Chiasmata

5

6

Joins called chiasmata form, and chromosome sections are swapped.

The new hybrid chromosomes move apart in the cell.

2 new nuclei are formed, each with a complement of 4 chromosomes.

CHROMOSOMES

The DNA of the cell nucleus is found in threadlike chromosomes, each made up of many fibrils placed side by side and complexly coiled. Each fibril comprises a pair of DNA molecules linked to proteins called histone. Other chromosome constituents include ribonucleic acid, RNA, one form of which transfers instructions from DNA to the cell cytoplasm. In most cells chromosomes are found in pairs, one member of each pair being derived from one parent, but chromosome numbers vary—human cells contain 23 pairs. Usually, chromosomes can only be seen distinctly when the cell is dividing, but a notable exception occurs in the fruit fly, *Drosophila*, whose giant chromosomes are clearly visible between divisions especially in their salivary gland cells.

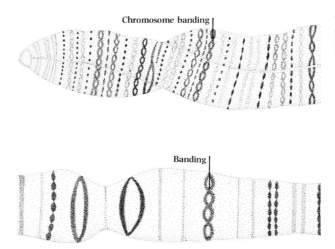

Chromosome banding

Banding

The giant chromosomes of the fruit fly, *Drosophila*, are boldly marked with bands. Experiments have helped to relate specific bands to the position of genes, the parts of a DNA molecule responsible for producing particular features in the adult fly. Although one band does not represent one gene but merely general gene positions, such studies do show that genes are arranged in line along the chromosomes.

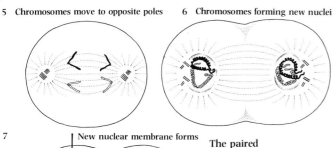

5 Chromosomes move to opposite poles   6 Chromosomes forming new nuclei

7   New nuclear membrane forms

iding cell membranes form

The paired chromatids draw apart at the centromere region and a member of each pair migrates to opposite poles. Nuclear and cell membranes form to create 2 new cells.

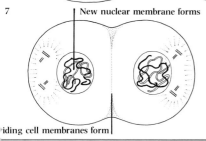

7   8

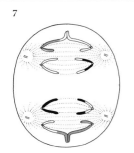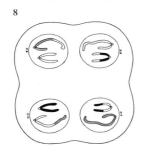

A second cell division now halves the chromosome number. The spindle-forming centrioles, similar to those formed in mitosis, divide again to form 4 poles. The chromatids separate and 4 new groups of chromosomes migrate, one to each pole. The nuclear and cell membranes reform, and 4 new cells, each with a single set of chromosomes, are ready for use.

9

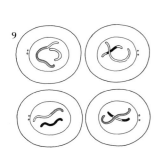

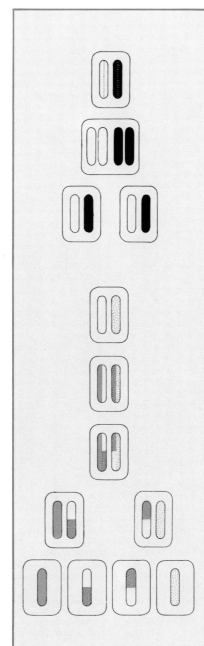

THE END-PRODUCTS

The ordered events of mitosis ensure that 2 cells are produced with identical complements of genetic material. Thus the 2 chromatids created as each chromosome is duplicated are exact copies. Mitosis also maintains the organism's status quo because it maintains the paired (diploid) arrangement of chromosomes common to nearly all body cells.

The 4 cells created as the end-products of meiosis each contain only a single representative of each chromosome pair in the original cell from which they came. But this is not the whole story. Because genetic material is exchanged by 'crossing over', as the paired chromatids separate in the first part of the division, the new chromosomes may contain novel gene sequences. These new arrangements, combined with the influx of material from another individual at fertilization, make for variation between members of the same species.

The start of new life

The job of the testes is to make male germ cells or gametes called sperm. At the moment of fertilization a sperm fuses with a female's egg to start the life of a new individual. The paired testes of a male mammal are made up of a mass of densely coiled tubes or seminiferous tubules which are enclosed within a fibrous bag or capsule. Covering the capsule is a sac of skin, the scrotum. The seminiferous tubules contain 2 types of cell: germ cells, which become sperm, and Sertoli cells, which provide the environment for sperm formation to take place. The proper production of sperm also depends on sex hormones. These are made by cells called Leydig cells, which lie in the spaces between the seminiferous tubules, under the control of the pituitary gland which is influenced by the brain.

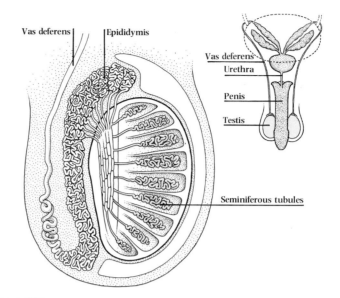

Inside the testis the seminiferous tubules connect with a network of small tubes (the rete testis) which run into the epididymis. The epididymis is so long —23 ft (7 m) in man—that it takes up to 14 days for sperm to pass along it. The end part of the epididymis is used for sperm storage. This is essential because although sperm are made continuously, ejaculation is occasional. From the epididymis sperm pass along the vas deferens and reach the outside of the body via the urethra.

While the testis produces huge numbers of sperm, the ovary of a mammal makes only small numbers of ova or eggs. Each of the ovaries of a human female is roughly the size and shape of an almond. The development of the eggs takes place in sacs or follicles, most of which are situated in the outer region of the ovary. Although a male makes no sperm until he is sexually mature, the eggs of a female are already partly developed when she is born. At this stage the eggs are called primary oocytes and have reached the first stage of the cell division (meiosis) that will eventually halve their number of chromosomes. Nothing more happens, however, until after puberty, when the oocytes complete their maturation under the influence of hormones made by the pituitary gland.

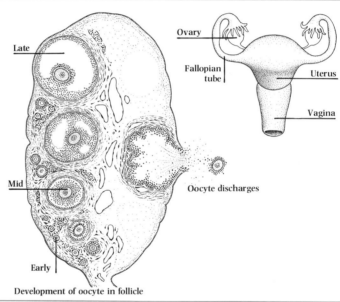

Development of oocyte in follicle

A regular cycle of changes occurs in the ovary. First, the cells around the oocyte in a follicle start to multiply. The oocyte begins to grow and a thick covering, the zona pellucida, is secreted round it. Under the influence of pituitary hormones the ovary wall ruptures and the egg is cast into the Fallopian tube. After the egg is shed the remains of the follicle are converted to a structure called the corpus luteum which makes hormones to help the uterus nourish the embryo until the placenta takes over.

The act of fertilization of an egg by a sperm takes place in the Fallopian tube of a female mammal. But to reach the ovum the sperm must push its way through the layers of material that surround the egg. These layers comprise the cells of the cumulus oophorus (egg-bearing little cloud) which remain round the egg when it bursts from the follicle and the zona pellucida. The sperm cannot complete the last—and most vital—stage of their long journey from the testis until they have spent a certain amount of time within the female. During this period, which lasts about 7 hours in humans and 2 or 3 hours in rats, the sperm undergo a process of activation called capacitation. Afterwards they are able to swim much more vigorously and can release the chemicals essential for penetrating the egg.

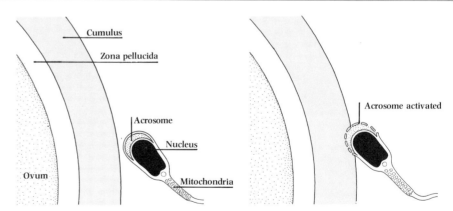

After several hours within the body of the female, sperm swim strongly toward the egg. Hundreds of sperm crowd round the egg, but only one will enter it, aided by enzymes at the head of the sperm and on the sperm body.

With the help of the enzyme hyaluronidase the sperm can burrow through the cells of the cumulus. As this happens pores appear in the membrane around the acrosome, but acrosome enzymes are often not discharged until the sperm reaches the zona pellucida.

Section of seminiferous tubules

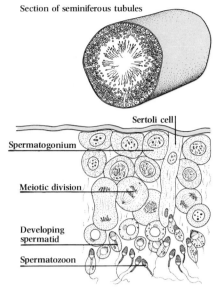

Sertoli cell

Spermatogonium

Meiotic division

Developing spermatid

Spermatozoon

In human semen sperm number about 100 million per ml. This is possible because the seminiferous tubules provide a vast area of cells from which sperm originate. In sperm formation the youngest cells are placed round the outside of the tube. As they mature into spermatogonia they are carried inward. During this passage their chromosome number is halved and they become spermatids. Once formed, spermatids become attached to Sertoli cells and mature into spermatozoa.

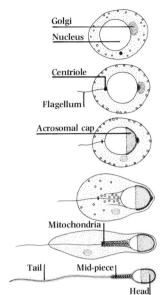

Golgi

Nucleus

Centriole

Flagellum

Acrosomal cap

Mitochondria

Tail

Mid-piece

Head

In the process of conversion from spermatid to spermatozoon one cell structure (a centriole) produces a tail or flagellum which endows swimming power. The Golgi apparatus of the cell, normally used for protein packaging, is converted into an acrosome, a body at the head of the cell, which contains chemicals vital to the penetration of an egg. The cell nucleus becomes condensed and elongated. The mitochondria, which provide energy for swimming, group round the flagellum.

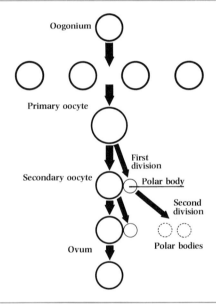

Oogonium

Primary oocyte

Secondary oocyte

First division

Polar body

Second division

Ovum

Polar bodies

Before birth, female germ cells, the primary oocytes, are formed from germinal stem cells in the ovary. The first of the cell divisions to halve the chromosome number usually takes place as the oocytes develop within the follicle. During this initial meiotic division the cell cytoplasm does not divide equally, and the smaller cell is rejected as a polar body. The larger cell or secondary oocyte is released from the ovary. Only if a sperm penetrates the egg is meiosis completed.

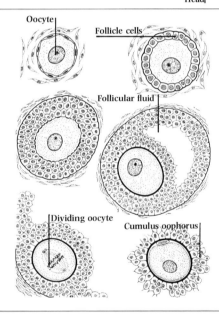

Oocyte

Follicle cells

Follicular fluid

Dividing oocyte

Cumulus oophorus

Changes take place in the follicle in which the oocyte is housed. As the oocyte grows, the cells of the follicle divide rapidly to enclose it. When the oocyte is about twice its original size the zona pellucida forms around it. Fluid is secreted into small pools between the follicle cells. The pools merge almost to fill the follicle. The oocyte is pushed to one side among a halo of cells—the cumulus oophorus. The structure is now a Graafian follicle and the egg is ready.

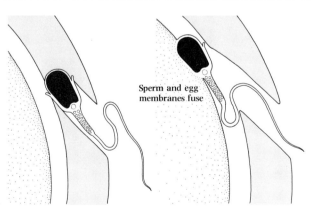

Sperm and egg membranes fuse

Sperm nucleus inside egg

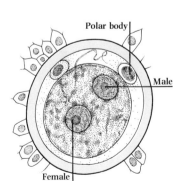

Pronuclei inside fertilized mammalian egg

Polar body

Male

Female

When the sperm reaches the edge of the zona pellucida the acrosome membrane is shed. The sperm first fuses on to the zona then makes a narrow slit in it. This is achieved by the action of the enzyme acrosin.

Another acrosome enzyme called neuraminidase also helps the sperm to penetrate and changes the zona pellucida chemically to prevent latecomers from gate-crashing their way into the egg. The sperm nucleus is now situated within the cytoplasm of the egg.

The nuclei of sperm and egg fuse together to make a cell with the full, diploid, complement of chromosomes—the same number as the parents from which they came. The fertilized egg now divides again and again to develop into a new individual.

The struggle to survive

Are animals friendly and cooperative in their dealings with one another? Or are they competitive and selfish individuals, primarily concerned with surviving long enough to reproduce and pass on their genetic characteristics to the next generation? A new study, sociobiology, which investigates the biological basis of the social behaviour of animals, takes the latter view.

In many ways these ideas are a return to Charles Darwin's theory of natural selection—that the struggle for existence ensures that only the fittest animals survive to reproduce. More recent views suggest that selection can act at the level of the group as well as the individual. This would encourage not only the formation of cooperative groups, but also perhaps the subjugation of individuals for the good of the group as a whole. When an animal in a group gives up an item of food or decreases its own chances of survival while increasing another animal's, this seems like altruistic behaviour. Ideas of altruism and group selection go together, and individual selection and selfish behaviour are similarly linked.

The problems of a group of sea birds breeding on an island demonstrate the difference between individual and group selection. If all the birds breed as soon as they can and raise as many young as can be fed, the population could increase until there are so many birds that they overfish the area and run out of food. Mass starvation would result and the colony might become extinct. If a nearby colony had a more sophisticated approach to population control and attempted to cut down the number of eggs laid, or made individual birds wait for several years to breed, they would maintain numbers suitable to their food supply and not suffer severe population crashes.

Sensible though it sounds, there is one major flaw in the group selection argument: all the individuals in the group must conform to the same rules. If one pair of birds was to breed earlier than the others or to lay more eggs, their offspring would inherit that tendency and these animals would soon spread through the population and replace the others. Group selection is always open to 'cheating' and before long individuals

are again competing with one another. This leads many to believe that evolution and selection must operate at the level of the individual—the ultimate goal of this individual being to leave behind as many young as possible therefore transmitting more genes to the next generation.

Taken to its logical extremes, the theory of sociobiology gives a picture of animals as survival machines, mere devices whose sole object is to ensure that their genes are passed on. The old saying that a chicken is just an egg's way of making another egg can be extended to the idea that an animal is a temporary device for ensuring the continued existence of its genes in another animal.

If this view is accepted totally then altruism becomes an impossibility. Examples such as the mother lion who gives up her food for her cub or the mother wildebeest who gives up her life, are explicable in that they are making sure that their genes survive in their sons and daughters. Even examples of cooperative behaviour where the animals are not related are not inci-

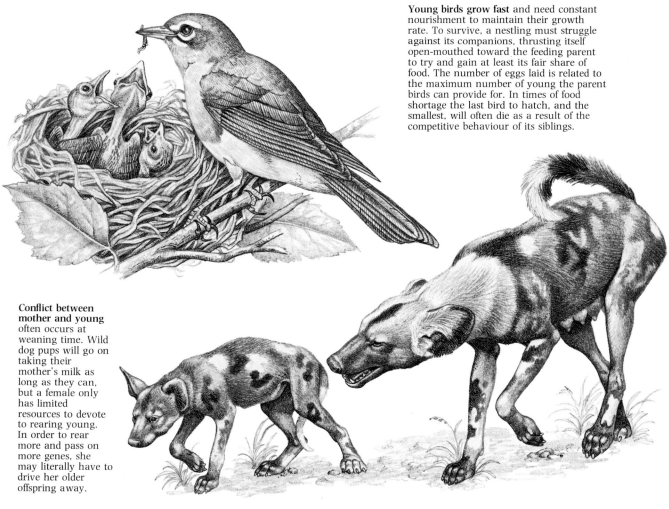

Young birds grow fast and need constant nourishment to maintain their growth rate. To survive, a nestling must struggle against its companions, thrusting itself open-mouthed toward the feeding parent to try and gain at least its fair share of food. The number of eggs laid is related to the maximum number of young the parent birds can provide for. In times of food shortage the last bird to hatch, and the smallest, will often die as a result of the competitive behaviour of its siblings.

Conflict between mother and young often occurs at weaning time. Wild dog pups will go on taking their mother's milk as long as they can, but a female only has limited resources to devote to rearing young. In order to rear more and pass on more genes, she may literally have to drive her older offspring away.

dences of true altruism. Rather than making any sacrifice the donor is taking out a form of insurance for the future. A baboon that helps a fellow baboon in a fight against a third, does so because of a learned relationship. If he finds himself in a similar situation the friend whom he helped before will come to his aid.

Animals can sometimes pass on more genes via close relatives than by reproducing themselves—an important factor in the lives of social insects such as bees. Worker bees guarding the hive entrance will sacrifice their own lives to save the hive. These workers are sterile and cannot reproduce; their best strategy is to protect the queen, their mother, and help her to produce more copies of their own genes.

In other types of animal close relatives take on extra importance because genes are shared between kin. An understanding of kin relationships explains why apparently altruistic behaviour is a common feature of group life. If a creature cannot help itself then it helps its young or other kin as they will all pass on some quantity of its genes. A male lion allows the cubs from his pride

to feed at a kill because he and his brothers have sired all those cubs and any of them may carry his genes. When male lions take over an existing pride they often kill off all the cubs. These cubs do not carry their genes and are therefore expendable.

The family itself, far from being a unit of total harmony, may be a selfish institution of struggle and strife. The amount of time and energy which each partner contributes toward the care of the young differs greatly in many mammals, particularly in polygamous species. While a male gets away with merely copulating, the female has a long period of pregnancy and suckling; she is literally left holding the baby. The male is left free to maximize his reproductive output and pass on even more genes by copulating with as many other females as possible.

Conflicts also occur between parents and their young; parents should raise as many young as possible to pass on more genes; so, although being fed suits the young and their future survival very well indeed, the mother will want to wean them and become pregnant again.

Often the female has physically to drive the young away before they will learn to live independently. There is also competition between the siblings in a litter or brood. If a youngster fails to compete he may well become the weakling or runt and eventually starve to death, particularly if food becomes short. Within a family group competition, aggression, exploitation and greed are all part of the relentless struggle to survive and reproduce at any cost to others.

There is no doubt that the theory of sociobiology presents a more ruthless and cynical view of animal behaviour than the traditional picture of a world of friendly, cooperative creatures. Its critics are particularly sensitive to any suggestions that human behaviour might be interpreted in sociobiological terms, and that some of man's behaviour could be rooted deep in his biological past. This idea is certainly a controversial one, particularly for those anthropologists, psychologists and sociologists who have largely ignored man's evolutionary history, but there are bound to be more developments on these new ideas in the future.

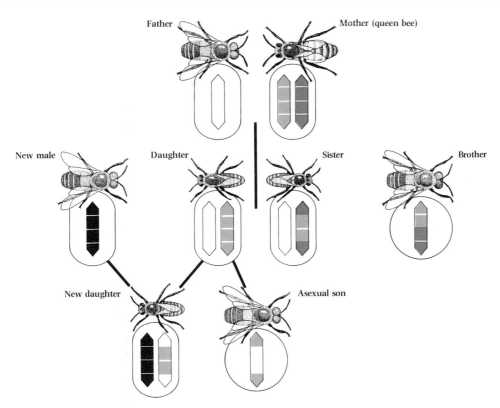

All worker bees are sterile sisters produced by the same queen. They cannot reproduce themselves, but by protecting the queen they ensure that more identical copies of their own genes are produced by her. If they did reproduce sexually their own young would actually contain fewer of their own genes than do their sisters.

Female bees have the normal 2 sets of chromosomes (diploid); males develop from unfertilized eggs and have only 1 set of chromosomes (haploid); a female honey bee receives half her genes from her diploid mother and half from her haploid father. Sisters have all the genes from the father as he only has 1 set, and share half of the

mother's. Mother and daughter are related by half whereas sisters share three-quarters of their genes. If a daughter mated with a new male she would be related to her new daughter by half. Thus the best way to pass on more genes is not to reproduce herself but to help her mother produce more sisters.

The act of fertilization

The union of sperm and egg at the moment of fertilization marks the start of a new animal life. Whether this union takes place inside the female body or outside it, sperm and egg must first be brought close together. For aquatic creatures the technicalities of getting eggs and sperm within fusing distance are easy—they simply shed their sex cells and the water carries them together—but even this process needs careful synchronization. Male and female must be mutually attracted when their sex cells are ready to be shed. In the vastness of the oceans the schooling behaviour of fish, and the homing of wide-ranging species such as salmon to spawning grounds, help to solve this distance problem.

Atlantic palolo worms, which live on the sea bed in the West Indies, have evolved a finely tuned coordination of egg and sperm release. In July, during the first or last quarter of the lunar cycle and at three or four in the morning, the rear segments of each worm's body, which contain eggs and sperm, break off and swim to the surface. At sunrise these open to release the sex cells, and the sperm fertilizes the eggs. On such a morning acres of the sea may be covered with fertilized eggs.

Another risk-reducing mechanism common to all animals in which fertilization is external is the vast over-production of eggs. A single female cod, for example, produces up to 7 million eggs, a turbot 10 million, a ling 28 million. Of these eggs only a small number are fertilized and even fewer reach maturity.

For animals that live on land, external fertilization is impossible because eggs and sperm cannot survive unprotected in the outside world. Terrestrial males thus introduce sperm direct into the reproductive tract of the female. Many land-dwelling invertebrate males present their sperm enclosed in a sac or package. This may be placed directly into the female with the aid of a special appendage or deposited on the ground. Sometimes elaborate manoeuvering is necessary to achieve sperm transfer. Males and females may be locked together for several hours while this is achieved. The male *Parnassius* butterfly even cements himself to the female with a special secretion during mating.

The penis is the evolutionary answer to many of the inherent difficulties of internal fertilization, for it is used to introduce sperm directly into the female. Many male insects, molluscs and vertebrates have penises but the penis is most sophisticated in mammals. The mammalian penis erects as its spongy tissues are engorged with blood during sexual stimulation. The penises of mammalian carnivores contain a bone, the os penis, which helps in the penetration of the female's vagina. This bone also increases vaginal stimulation which is an important trigger for the release of mature eggs in many of these animals.

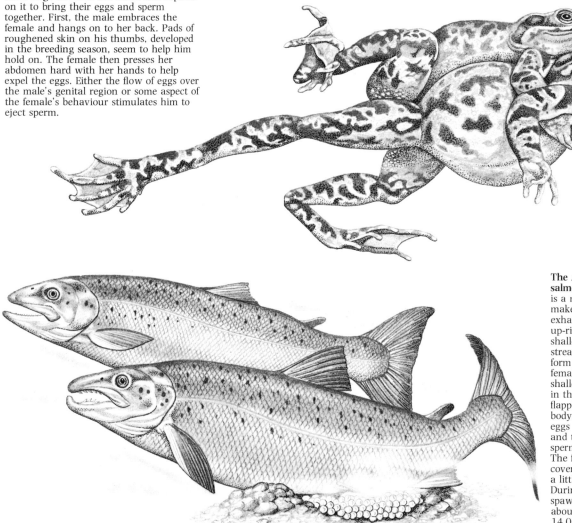

Most frogs breed in the water and depend on it to bring their eggs and sperm together. First, the male embraces the female and hangs on to her back. Pads of roughened skin on his thumbs, developed in the breeding season, seem to help him hold on. The female then presses her abdomen hard with her hands to help expel the eggs. Either the flow of eggs over the male's genital region or some aspect of the female's behaviour stimulates him to eject sperm.

The Atlantic salmon, *Salmo salar,* is a marine fish but makes an exhausting journey up-river to spawn in shallow freshwater streams. The fish form pairs and the female scoops out a shallow depression in the river bed by flapping her tail and body. She releases eggs into the nest and the male sheds sperm among them. The female then covers the eggs with a little gravel. During the spawning time of about a fortnight 14,000 eggs may be laid.

**The male and
female scorpions**
face each other,
abdomens held high
in the air. He seizes
the female and they
walk backward and
forward in a strange
dance that may last
for hours, even
days. Eventually the
male deposits a
small packet of
sperm on the
ground and leads
the female into such
a position that a
hook on the sperm
packet attaches to
her genital opening.
The sperm fertilizes
her eggs internally.

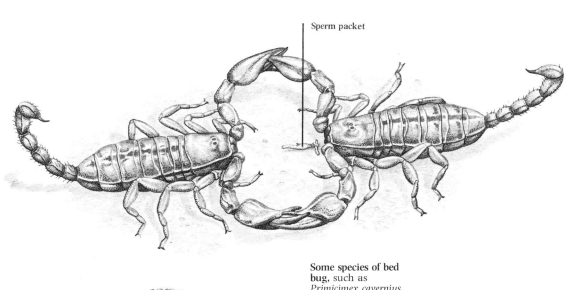

Sperm packet

**Some species of bed
bug,** such as
Primicimex cavernius,
inject sperm directly
through the outer
covering of the
female's abdomen.
The sperm is
dispersed by the
blood to all parts of
her body.

Hectocotylus

The third right arm,
the hectocotylus, of
a male octopus is
modified for placing
sperm into the
female. With this
arm the octopus
takes a mass of
sperm packages
from a storage sac
and puts them into
her mantle cavity.

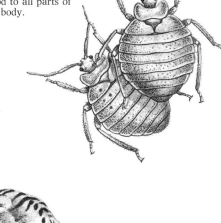

**A female tiger on heat becomes restless
and active** and gives mating calls to attract
males. If more than one male appears they
fight and the loser is driven away. The
male and female tigers, *Panthera tigris,*
approach each other and gently kiss,
allowing whisker to whisker contact. The
female behaves playfully and rolls on her
back. She gently nips the male and
presents herself to him. He mounts her and
inserts his penis into her vagina. At the
climax he grips the female's neck and gives
a piercing scream. Copulation lasts only a
few minutes but is repeated 3 to 20 times
a day for up to 3 weeks.

Male and female attire

Nature dresses the males and females of many animal species in different costumes suited to the roles each performs in the drama of sexual life. Males and females often differ in both appearance and behaviour—a phenomenon known as sexual dimorphism.

Where males and females are dissimilar in simple ways, as in colour or markings, these differences aid sexual reproduction by making potential mates easy to recognize. Even if male and female look alike, they may differ essentially in other ways. Thus females may produce special odours, while the whole life-style of the male is often quite unlike that of the female. While he is aggressive, she is submissive and plays a more direct part in the care of the young.

During a breeding season, aggressive males challenge each other for territory and to achieve dominance over other males. The victors then attract females to mate with them. If males must dispute to become privileged breeders, they tend to be larger and stronger than females and may also have anatomical weaponry such as tusks or antlers to assist them in the fight. In many species this inter-male rivalry is a mock battle of ritual displays without physical contact. To make these displays, males often have special body markings and flamboyant colours. The females, who are mere spectators, appear dowdy by comparison.

The bodily characteristics that promote success in fighting to establish and defend a territory are particularly marked when polygamy is practised among males. Polygamy is most likely to evolve in habitats that restrict breeding territories to a few males. Having become dominant, a male promotes the survival chances of his complement of genetic material by mating with a whole harem of females. In deer, for example, evolution has favoured those males whose genetic make-up leads to the annual shedding and regrowth of the stag's antlers. This cycle of change is advantageous because antler damage can seriously impair a male's ability to joust with his rivals. The extinct giant elk, *Megaloceros*, had massive antlers weighing more than its entire internal skeleton; these must have represented a huge investment of metabolic resources.

Males are not always larger than females. In some fish—angler fish being the prime example—the male is not only much smaller than his mate but lives permanently attached to her as a metabolic parasite. This extreme marriage assures the association of the sexes in the open acres of the ocean where more conventional relationships might not succeed.

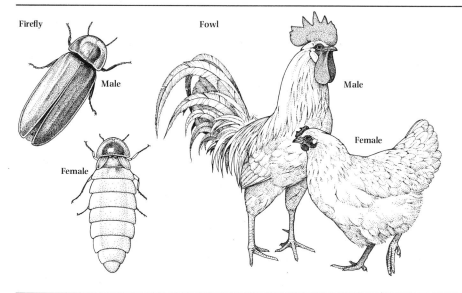

Firefly
Male
Female

Fowl
Male
Female

Peacock
Female
Male

Baboon
Male
Female

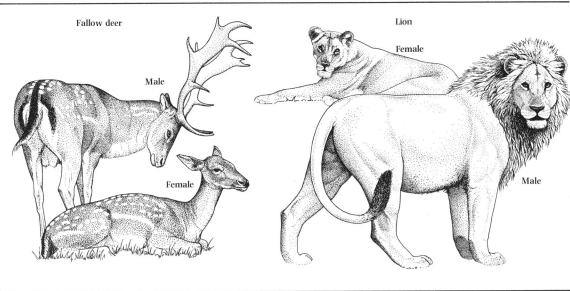

Fallow deer

Male

Female

Lion

Female

Male

EXTERNAL FEATURES

Sexual differences are often obvious characteristics. The female firefly is wingless whereas the male has wings. The male of the domestic fowl has a distinctive red comb dependent on hormones produced by the testes. The impressive antlers of the fallow deer stag are shed each autumn and regrown for the breeding season. As lions live in prides with a group of females and juveniles, but only a small number of adult males, the mane is probably an indication of status.

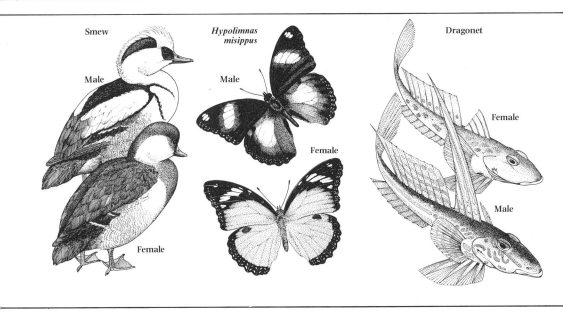

Smew

Male

Female

Hypolimnas misippus

Male

Female

Dragonet

Female

Male

PLUMAGE AND MARKINGS

The peacock is a native of India. The male has magnificent plumage and keeps a harem of 2 to 5 females. The plumage of the male smew is again more distinctive than that of the female. In the butterfly, *Hypolimnas misippus*, the male has dark brown and vivid blue markings; the plainer female is brown and white. The yellowish-brown male dragonet fish has bright blue markings and the first spine of his dorsal fin is elongated.

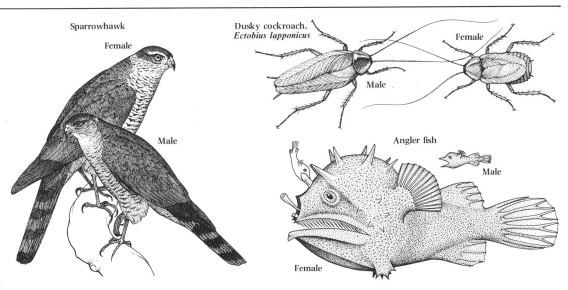

Sparrowhawk

Female

Male

Dusky cockroach, *Ectobius lapponicus*

Female

Male

Angler fish

Male

Female

SIZE DIFFERENCE

Male baboons are roughly twice the size of females and sport a large mane. They are aggressive toward each other. In many birds of prey size difference is reversed—the female sparrowhawk is up to a third larger than the male. The male cockroach is about a third larger than the female and quite distinct in appearance. The contrast between the male and female angler fish is extreme. The male is up to 6 in (15 cm) long and lives as a parasite on the 3 ft (90 cm) long female.

Blueprint for the body

Throughout the animal kingdom there are several formulae for the determination of sex. Thus some animals are hermaphrodite, with both male and female sex organs, some change sex in the course of their development but most have two distinct sexual forms, the male and the female.

The clue to the determination of sex lies in the hereditary blueprint carried in all body cells in the chromosomes of the nucleus. The chromosomes contain the building instructions for the whole of an animal's body—including its sex. Long before the discovery of DNA, chromosomes provided the key to the way in which sex is determined. In 1902 the sperm of a species of bug were found to be of two types, one with an extra chromosome. Eggs fertilized by sperm carrying the extra chromosome always developed into females while those fusing with sperm without this chromosome became males.

Since this first breakthrough, a link between chromosomes and sex determination has been firmly established for a great many—though by no means all—animals with separate males and females of the species. It is nearly always the pattern of chromosomes in the fertilized egg that seals the sexual fate of the new individual arising from it.

To distinguish them from the autosomes, the chromosomes not directly involved in determining sex, the sex chromosomes are usually given letter labels. Take the case of the bugs, whose sex-determination mechanism is shared by several other insect species. Each egg of the female bug contains one unpaired (haploid) set of autosomes, plus one sex-determining X chromosome. The sperm also carry a single set of autosomes and may have one X chromosome or none, a condition labelled as O.

At fertilization, the two sets of autosomes from sperm and egg combine to restore the normal, paired (diploid) chromosome number. The X chromosome of the egg may be paired with an X from a sperm and develop into a female, but if the XO combination occurs the offspring will be male. This pattern of events is the XX-XO system.

Many animals have two sets of sex-determining chromosomes. In the XX-XY pattern of mammals, most frogs, some fish and true, two-winged flies, all eggs carry an X chromosome, the sperm an X chromosome or a Y chromosome which is smaller than the X. The XX combination at fertilization produces a female, XY a male. In birds, most reptiles, some amphibia, fish and insects there are also two sorts of sex chromosome, but in a different conformation. The males of these creatures produce sperm with only one kind of sex chromosome, the Z, while females make eggs that may have a Z or a W chromosome. In contrast to the XX-XY system, the pairing of equals (ZZ) produces male offspring.

Honeybees are a notable exception to the sex chromosome rule, for no special chromosomes are involved. The females have the full, diploid chromosome number while the males, which develop asexually from unfertilized eggs, have the halved, haploid chromosome complement. For males to produce sex cells—which must also be haploid—a special sort of cell division is needed that 'shuffles' the genetic material but avoids halving the chromosome number.

The roles of the X and Y chromosomes and their Z and W equivalents depend, as for all chromosomes, on the genetic material they carry. Experiments have shown that only about half of the genetic material of a Y chromosome is vital

The chimpanzee, *Pan troglodytes,* has 48 chromosomes: 23 pairs of autosomes and 2 sex chromosomes. Other apes, such as the orang-utan and gorilla, have the same number. Man has 22 pairs of autosomes and 2 sex chromosomes; it may be that at the early stages of his evolution he had 23 pairs but 2 fused to give the present number.

Chimpanzee

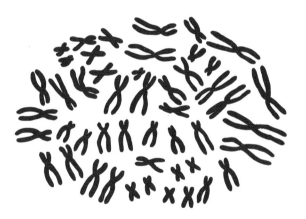

Bird chromosomes are of 2 kinds: large macrochromosomes and small microchromosomes. The total number is difficult to assess. Male birds have only 1 kind of sex chromosome and are ZZ. Females are thought to have 2 different sex chromosomes (ZW). The W chromosome has been positively identified in the budgerigar, but the possibility remains that some females have a ZO pattern.

Fowl, *Gallus domesticus*

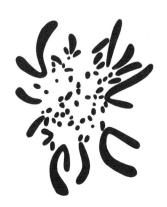

to its sex-determining function—the rest is 'padding' to make the chromosome large enough to match up to the physical demands made on it.

Of the two X chromosomes that go to make up a female, only one is genetically active in producing femaleness. The other is 'knocked out' by the cell and becomes a 'Barr body' which lies close to the membrane surrounding the nucleus. This inactivity of one X chromosome takes place early on in development. In some cells of mammals the X from the mother continues to function, in others the X from the father.

The X and Z chromosomes are larger than the Y and W chromosomes. Although they are vital to the determination of sex, X and Z chromosomes also carry genes for body characteristics that have nothing to do with sex. But the inheritance of these characteristics is linked with sex simply because the genes are on the sex chromosome. Colour blindness and haemophilia in humans are good examples. The genes for these disorders are carried on the X chromosome.

A woman with red-green colour blindness in the family, but who does not herself suffer from the defect, may carry the colour blindness gene on one of her X chromosomes. Her colour vision is not affected because this gene is masked (dominated) by the normal gene on her other X. The eggs she produces will be of two kinds, one with and one without the colour blindness gene. If the egg carrying the deleterious gene is fertilized by a Y-carrying sperm her son will be colour blind because the gene is no longer masked by a normal one.

Like all genes, those involved in sex determination and carried on the sex chromosomes issue instructions to direct the activities of body cells, but the way in which they accomplish this task is far from clear. Fertilized animal cells have the inbuilt potential to develop into either sex—the sex chromosomes merely tip the balance in one particular direction. In vertebrates it seems certain that the sex chromosomes devote their efforts to ensuring the development of ovaries and testes. Once these organs are formed, they release hormones which govern the formation of the other main sex organs, the secondary sexual characteristics such as breasts and beards, and the behaviour patterns vital to successful courtship and mating.

Even in vertebrates the sex chromo-somes may be thrown off course. In amphibia, for example, larvae exposed to low temperatures in early life develop into females, while high temperatures produce males.

Apart from the sex chromosomes, the other chromosomes, the autosomes, may also play their part in sex determination. In the fruit fly, *Drosophila*, irregularities in the cell divisions producing eggs and sperm may mean that a fly is produced with an extra set of autosomes. If two X chromosomes are also present in each cell the biological norm is upset and the adult is not a female but intermediate in sex with some male features.

It has been suggested that the disturbance of sex balance in some humans may be due to similar circumstances involving the autosomes, but nothing has yet been proved. Certainly humans are not immune from errors as far as their sex chromosomes are concerned. Boys may be born with XXY, XYY or XXYY chromosome conformations, for example, while girls may be XO, XXX, XXXX or XXXXX. The extra or missing chromosomes do, unfortunately, cause varying degrees of physical and mental impairment.

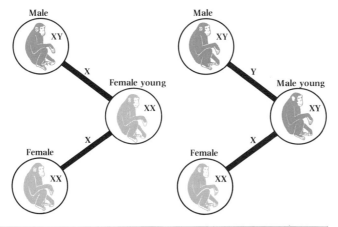

Each sperm of a chimpanzee contains 23 autosomes and an X or a Y sex chromosome. The female ovum, or egg, must have an X sex chromosome as the female only carries X chromosomes. If a sperm carrying the Y chromosome fertilizes the egg the offspring will be male. If the sperm carries the X chromosome the offspring will be XX and female.

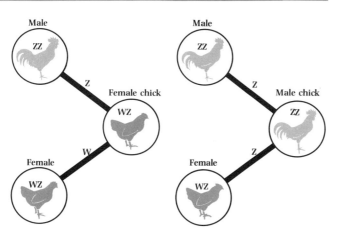

The sperm of a bird must contain a Z chromosome. An egg may contain a Z chromosome or a W (in some species perhaps O). Thus, in contrast to mammals, the female's egg is the sex-determining factor in birds. If an egg carrying a Z chromosome is fertilized, the offspring will be male; if the egg carries a W (or O) chromosome, it will be female.

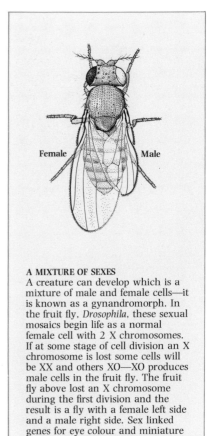

A MIXTURE OF SEXES
A creature can develop which is a mixture of male and female cells—it is known as a gynandromorph. In the fruit fly, *Drosophila*, these sexual mosaics begin life as a normal female cell with 2 X chromosomes. If at some stage of cell division an X chromosome is lost some cells will be XX and others XO—XO produces male cells in the fruit fly. The fruit fly above lost an X chromosome during the first division and the result is a fly with a female left side and a male right side. Sex linked genes for eye colour and miniature wings are also involved.

Changing sex

The separation of the sexes into male and female is for most animals—and especially for insects and vertebrates—one of nature's life insurance policies. It ensures the creation of a new generation by equipping male and female to perform the specialized tasks that reproduction demands. But this is not a universal condition; the sexual make-up of animals can vary, and many invertebrate animals have no separate sexes at all.

One problem, however, of separate sexes is that only encounters between males and females can lead to mating. Flukes and flatworms, earthworms and ragworms, slugs and snails are among the many invertebrate creatures that have no male–female distinction. Each individual is hermaphrodite, producing both sperm and eggs. It is possible, in theory, for an hermaphrodite to fertilize itself, but this is rare in practice. Usually a variety of anatomical and behavioural devices prevents self-fertilization and

ensures sperm swapping between two individuals. The advantage of the hermaphrodite way of life is that any two sexually mature individuals can mate.

Even when there are separate male and female forms, sexuality may vary. In nearly all animals the ratio of males to females is 50:50—perhaps surprising since a single male can fertilize many females. The ratio is normally maintained by the sex chromosomes, but some animals have evolved mechanisms of overriding the instructions the chromosomes carry to aid their reproductive effort.

The environment is the key to many such mechanisms. Any animal that lives a sedentary existence has a mating problem. How can it ensure that males and females settle close together? Slipper limpets do so by settling on top of one another and changing sex according to their position in the pile, thus guaranteeing that the colony always contains males and females. In the marine worm,

Ophryotrocha puerilis, the overriding factors in sex determination are age and food. Young worms are always male but, as they grow, they change into females. These females may revert to maleness if they are deprived of food.

Another sea-dwelling echiuroid worm, *Bonellia viridis*, shows how the environment can influence sex. The female of this worm is several inches long, the male a microscopic parasite that lives attached to her reproductive tract. The sex of larvae that hatch from the fertilized eggs is decided by their lifestyle. Free-swimming larvae develop into females but those that develop in contact with the extended mouth (proboscis) of a female become males.

Among vertebrates the environment only has a hand in sex change in certain fish. Those of the families Labridae, Sparidae and Serranidae can alter from male to female or vice versa, switching back and forth in response to reverse changes in a partner.

A FISH HAREM
The male cleaner fish, *Labroides dimidiatus*, leads a harem of females. If the male leader dies, the dominant female in the harem changes sex and takes over. Within a few hours she/he begins to make male aggression displays; courtship and spawning behaviour follow in a few days.

Dominant female   New leader

Male

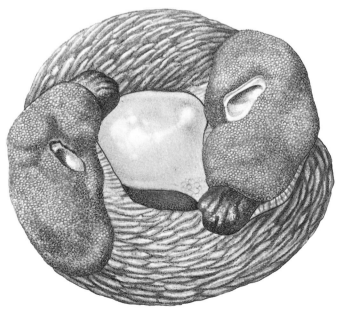

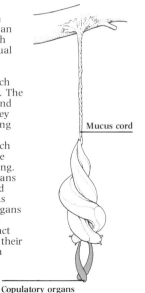

Slugs and snails are hermaphrodite—a combination of both sexes. They have an organ, the ovotestis, that produces both eggs and sperm but have separate sexual ducts and male and female copulatory organs. Usually two animals meet to exchange sperm so that the eggs of each are fertilized by the sperm of the other. The slug, *Limax maximus*, finds a partner and both climb up on to a wall or tree. They circle for about 30 to 90 minutes licking and covering each other with mucus. Quite suddenly they entwine and launch themselves into the air. As they fall the mucus forms a rope on which they hang. Both slugs extend their copulatory organs and wrap them together—the extended organs are about 4 in (10 cm) long. As the sperm exchange takes place the organs change shape to look rather like small umbrellas. Afterwards the organs retract carrying the sperm. The slugs can eat their way back up the mucus cord although often one will drop to the ground.

Mucus cord

Copulatory organs

The sex of the male slipper limpet can change if his position in the colony demands it. The male organs pass through a transitional phase when the penis is reduced in size and finally disappears. An oviduct may develop to complete the change to a female.

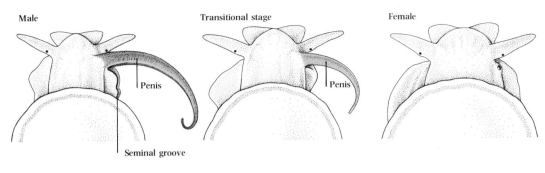

Penis

Penis

Seminal groove

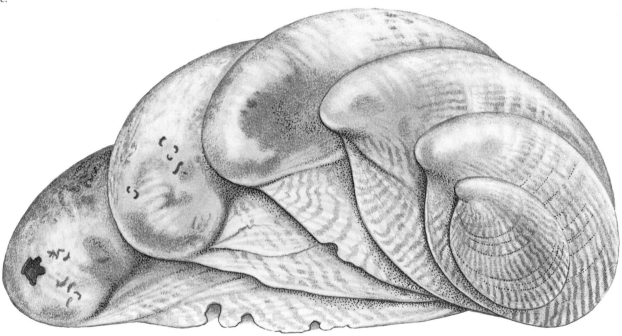

Slipper limpets live in colonies of animals piled up on top of one another. A colony starts when a free-swimming larva of the slipper limpet, *Crepidula fornicata*, settles on a rock to start its adult life. More larvae settle on top of the first. The older limpets at the base are female, and the young, more recent, settlers at the top are male. As a male ages he goes through a transitional period during which his male organs degenerate. The limpet can then re-develop into a male or change into a female, depending on its position in the pile. A male will remain male as long as he is attached to a female.

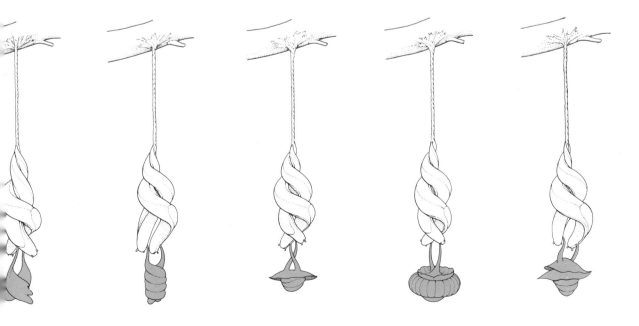

Hormone messengers

For an orchestra to perform in perfect harmony, the playing of each instrument must be accurately synchronized. So it is with sexual reproduction, the success of which demands the coordination of many complex processes. This coordination ensures the maturation, meeting, mating and procreation of males and females.

Except in the most simple of animals, the joint action of all tissues and organs is achieved in two ways. Immediate and short-term adjustments are mostly mediated by the nerves, which are capable of rapid information transfer. Slow, rhythmic, long-term or even permanent changes in function or behaviour are effected by chemicals carried in the blood. These chemical messengers are hormones, and the tissues that make and release them are the ductless or endocrine glands.

Earthworms, crabs, octopuses and many other invertebrate animals rely on hormones to regulate the development of males and females and to underwrite such vital processes as egg laying. But it is in vertebrates, and particularly in mammals, the group of which man is a member, that hormones can play the most sophisticated role in sexual reproduction.

In all vertebrates the sex glands or gonads—ovaries in the female and testes in the male—not only produce sex cells but also act as hormonal glands. Yet these glands are not masters of their own destiny. They cannot operate until directed by the pituitary gland.

Situated just beneath the brain, the pituitary produces hormones which stimulate many other ductless glands into activity, among them the sex glands. And the siting of the pituitary is no anatomical accident, for its hormone output is regulated by the hypothalamus, the part of the brain to which it is attached and which provides a link between brain and internal organs.

Sex hormones may start their work even before an animal is born. In mammals, a developing embryo has the innate capacity to become either male or female. In the male, hormones called androgens are released which trigger tissues to differentiate into specifically male organs, among them the penis. The female pattern, giving rise to such characteristics as the Fallopian tubes and the womb or uterus to house a developing embryo, seems to be built into the hereditary blueprint and does not demand hormonal intervention.

During its juvenile years, an animal's sex glands produce only low levels of hormones. With the onset of adulthood, a stage known in humans as puberty, the sex glands, prompted by secretions from the pituitary, start to produce their hormones. In the female these hormones are estrogens and progestogens, of which the most important are estradiol and progesterone. In the same way the male sex glands release several androgens, of which testosterone is the most influential.

In response to higher hormone levels, juvenile animals begin to look and behave like adults. The internal sex organs grow larger, the testes start to produce sperm, the ovaries to release mature eggs. The adjuncts to sexual maturity—the secondary sexual characteristics—also develop now. The

The pituitary gland releases several hormones which influence the hormone-secreting activities of other glands. One of these stimulates the adrenal gland, another the thyroid gland, and others the sex organs. The pituitary gland itself is controlled by the hypothalamus of the brain. Hypothalamic-pituitary activity is subject to both environmental and internal physiological influences.

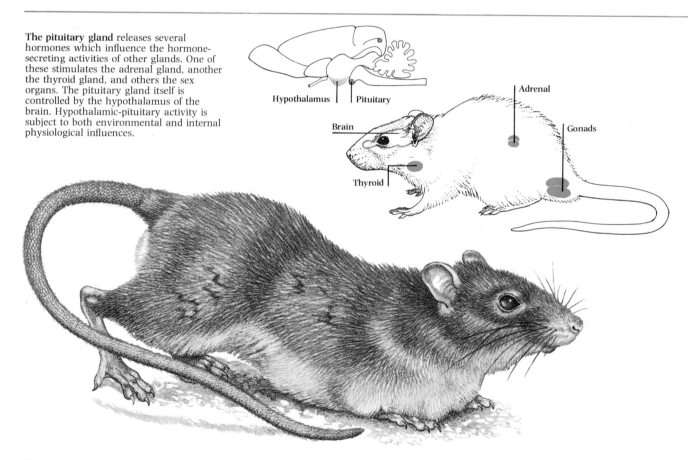

The sexual reactions of a female rat are dependent on her ovarian hormones. Only during the estrus or 'in heat' phase, when the estrogen levels in the blood are high, does she show interest in mating. In this receptive state she is more active than usual; she will approach a male rat, perhaps sniff at his genitals and make short hopping movements. The female then takes up a sexual presentation posture: she curves her back and moves her tail aside to expose her genitals, then stands rigid so as to withstand the weight and thrusting action of the male. This posture is prompted by her hormones.

deer's antlers, the lion's mane and the beard of a man are all products of testosterone influence in males. In female animals the sex hormones prompt the development of the breasts or mammary glands.

The release of mature cells and female sex cells is essential if the 'contract' of mating is to be fulfilled in the form of a new individual. In female mammals, eggs are released one or a few at a time in a regular cycle. Human females are no exception. At puberty, a girl's ovaries contain some 250,000 immature eggs or ova. Of these only about 400 ever leave.

Each month a regular cycle of hormonal events occurs. The pituitary first releases a follicle stimulating hormone (FSH) which triggers the final maturation of one egg and causes its follicle to secrete the hormone estradiol. Carried to the uterus in the blood, estradiol causes a buildup of the uterus, thus preparing it to receive and nurture a fertilized egg. After about 14 days, more FSH plus a sudden surge of luteinizing hormone (LH) come from the pituitary.

This pulse of hormones makes the follicle rupture and relinquish its mature ovum into the Fallopian tube linking ovary and uterus. Fertilization may take place in the tube or the ovary.

The now empty follicle does not disintegrate at once but, under LH influence, is converted to a structure called the corpus luteum which makes progesterone and some estradiol. If the ovum is not fertilized the corpus luteum begins to degenerate and lower hormone levels cause the lining of the uterus to be shed as the menstrual period. But if fertilization does take place the corpus luteum is maintained by another hormone, chorionic gonadotrophin, made by the placenta in the uterus wall.

Across the mammalian spectrum, shades of difference occur in the female pattern. Cats, rabbits and ferrets, for example, do not release eggs (ovulate) automatically—for them the surge of LH depends on the physical stimulation of copulation. Man and his primate relatives are the only mammals to have menstruation. In others the uterus

lining is absorbed. When this happens the cycle is called estrous, and animals 'primed' with high hormone levels just before ovulation are described as in estrus or 'in heat'. And while human females release one egg each month, many mammals ovulate only at a time corresponding to the best season for the birth of the young.

The male pattern for mammals is much simpler. The same pituitary hormones are pumped out regularly (on a day-to-day basis) to maintain the production of mature sperm from the testes.

However efficient the sperm and egg manufacturing machinery, it is useless unless combined with the appropriate sexual behaviour. Again the hormones made by the sex glands are involved. Non-primate females will only accept the sexual advances of males when they are in heat. Primate females, including humans, are emancipated from the dictates of their menstrual cycle and, given the right handling, will mate at any time. For them, sexual motivation is controlled by hormone output from another endocrine gland, the adrenal.

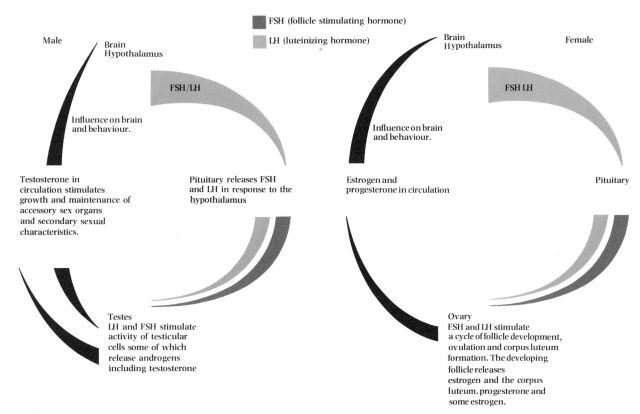

A male's pituitary gland produces FSH and LH. LH stimulates the Leydig cells in the testes to produce androgens, mainly testosterone, and FSH stimulates activity of the Sertoli cells in the testes. Both types of cell are necessary for sperm production. Testosterone also stimulates the accessory sex glands and the development of male anatomy. It influences the brain in promoting male sexual behaviour.

A female's pituitary produces FSH and LH, but in a cyclical manner. FSH initiates an ovarian cycle. The ovarian hormones, estrogen and progesterone, also influence sexual anatomy and behaviour.

Reproductive organs

Bony fish vary greatly in their breeding mechanics. As a rule fertilization is external and the eggs laid by the female are either floaters or sinkers. Because floating eggs are prey to more predators, more are laid. The ling, with floating eggs, lays up to 28 million; the herring, whose eggs sink, only 500,000. Another factor that reduces the need to lay large numbers of eggs is parental care, such as the attention sticklebacks give to their eggs. From external appearances, male and female fish may be difficult to differentiate, but many males take on bright nuptial colours in the breeding season.

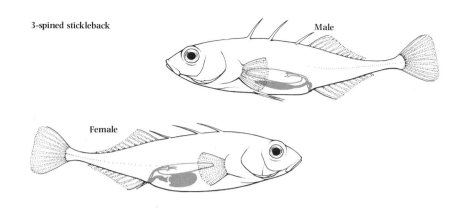

3-spined stickleback
Male
Female

Although solitary for most of the year, frogs and other amphibians breed communally. In spring, male frogs precede the females to the water and croak to attract a mate. In copulation the male clings to the female with a thickened pad of skin on each hand. The mounted female lays 500 to 5,000 jelly-clad eggs which as they emerge are fertilized outside her body by the male's sperm. Egg laying is completed in only about 10 minutes. In some frogs fertilization takes place inside the female, but the only one to give birth via the reproductive opening is *Nectophyrnoides* of Africa.

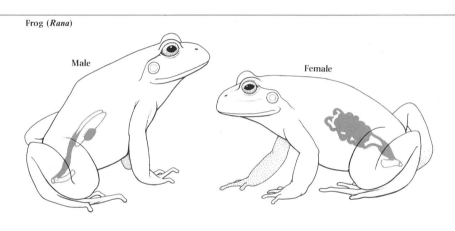

Frog (*Rana*)
Male
Female

Male and female birds are often easy to tell apart, for the male is generally clad in more brightly coloured plumage. After an elaborate courtship involving song and nest-building, the pair mate. Sperm are deposited inside the female and her eggs fertilized internally. The large yolk-filled, shell-covered eggs necessitate large-scale rearrangements of the female bird's sex organs. To make enough space inside her body only the left ovary develops; the right one remains rudimentary. The number of eggs per clutch varies from 1 or 2 in some hawks to 18 to 20 in quails.

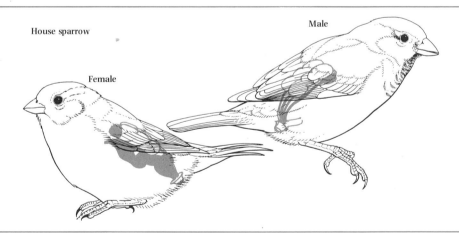

House sparrow
Male
Female

The reproductive anatomy of a rabbit is representative of the mammalian group. Most mammals have definite breeding seasons when the female's ovaries release up to 10 mature eggs. Fertilization is always internal and, except in the most primitive mammals, the embryo is nurtured in the female's uterus by means of a placenta. Male mammals may have noticeable body ornaments, such as manes or antlers. Most have testes slung outside the body (because sperm production demands cooler conditions than those inside the body) and a penis that erects when its spongy tissues fill with blood.

Rabbit
Male
Female

Male bony fish are unusual because they have not adapted a tube from the kidney for sperm transport, although, as in all vertebrates, their testes and kidneys arise side by side in the embryo. Instead they have sperm ducts, made from testis tissue, which lead to the cloaca behind the anus, where the kidney ducts also discharge their urine. There is no penis.

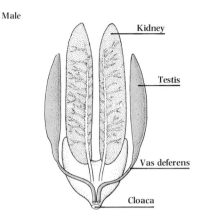

The vast number of eggs made by female bony fish are produced in ovaries that are not directly joined to the oviducts that carry eggs to the cloaca. Instead, eggs released by the ovaries are caught by a funnel at the top of each oviduct. To prevent eggs from escaping and filling the body cavity, ovary and oviduct are enclosed in a membrane sheath.

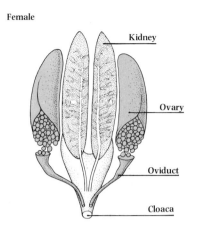

Sperm from the male frog's testes travel through a tube that was previously connected to the kidney. During development, testes and kidneys share a close association and a new tube arises for urine transport. All the systems have a common exit at the cloaca. Fertilization is external so there is no penis.

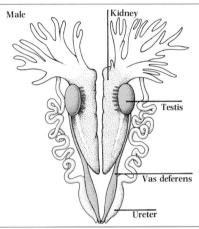

The eggs of the female frog are produced in lobed ovaries. Mature eggs pass down the oviduct where they acquire a jellylike covering from secretions of the oviduct wall. A sac at the end of each oviduct acts as an egg store. Urine, which travels from the kidney, eggs and digestive wastes, are all released through the cloaca.

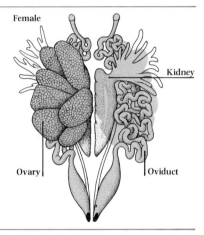

In the breeding season a male bird's testes may enlarge over 500 times. Sperm leave the testes via the epididymis and vas deferens. The latter is also used for sperm storage as are the seminal sacs, if present. Although some birds do have a well-developed erectile phallus, in most there is simply a small thickening of the cloaca.

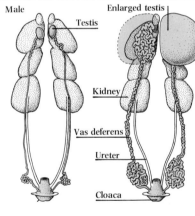

After mating, sperm are stored in the infundibulum and vagina of the female bird, then swim to the yolky eggs in the single, left ovary. Fertilized eggs then pass down the oviduct developing as they go. Egg white is added in the magnum of the oviduct, the shell membranes in the isthmus, and the shell in the uterus. Eggs are expelled via the cloaca.

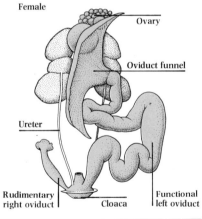

On their journey from testes to penis the sperm of male mammals pass through the epididymis and the vas deferens (which also act as sperm stores) and then into the urethra. Substances secreted by the seminal vesicles and prostate gland are additions essential to sperm potency. Although urine also leaves the body via the urethra the anus is separate.

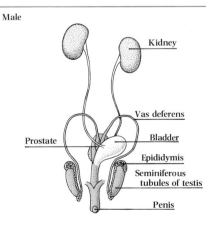

As an egg leaves the ovary of a female mammal it is caught in the infundibulum of the Fallopian tube. Fertilization takes place high up in the tube; the fertilized egg passes to the uterus. Next, the developing embryo buries itself in the uterus lining and a placenta develops. In primates and some rodents the vagina lengthens to accommodate the penis during mating.

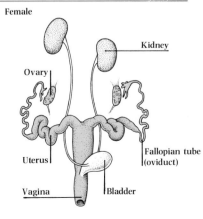

Growing up

The time it takes for an animal to reach sexual maturity depends on two factors: the size of the creature and its span of life. Small animals rarely live long. They have opted for a strategy of rapid reproduction and hence, quick colonization of new habitats. Their size is linked to these abilities. Whether their genes produce small, unstable survival machines is immaterial. What is vital is that the hereditary material is carried on from one generation to the next.

The other extreme of this situation is to leave reproduction until late in life and become a large animal. Many creatures fall between these two extremes.

Large size, long life and delayed reproductive maturity tend to go hand in hand. Size alone has specific advantages. A large animal is able to keep its body temperature stable with greater ease than a small one, and size serves as a protection against predators. But although size gives superiority in the struggle for survival, it takes both time and a stable, predictable environment for a young animal to grow. In most animals, reproductive maturity is delayed until growth is complete, for only fully grown males can compete successfully for mates, and only full-size females

have the internal resources to produce offspring. In mammals, and particularly the carnivores and primates, a long juvenile period is essential for learning the skills necessary for adult life.

Because the young of large animals cannot achieve sexual maturity quickly, they have to withstand a long period of vulnerability if they are to become parents themselves. A young, isolated buffalo is easy prey to lions, but equally, a juvenile lion would quickly starve without the care and protection of its elders. In such circumstances a successful reproductive strategy is to give birth to only a few offspring but to make a great investment in their survival.

Life in a highly varied and harsh environment can also make delayed breeding and small numbers of well protected young a necessity. Eagles, condors and albatrosses depend for survival on food which is sparse and hard to find. Usually only one chick is produced at a time, the male doing all the hunting for himself and his family. Condors and royal albatrosses do not begin to breed until they are about eight years old, and a pair of crowned eagles will feed and protect a single offspring for at least a year and a half before it is fully fledged.

Weeks

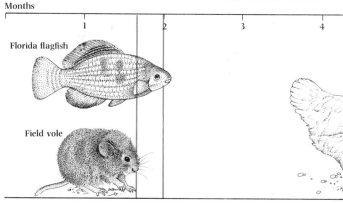

Months

The field vole is a small mammal weighing up to 1 oz (28 g), which reaches maturity in a couple of months. The Florida flagfish takes 60 days. The flour beetle is mature in about 6 months. A number of males inseminate a female, but the most recent sperm to enter her sperm store usually fertilize her eggs.

Florida flagfish

Field vole

Mayflies live as aquatic nymphs for between 1 and 3 years. Once they emerge as winged adults they live only a few days, or even hours in some species. Bears mature slowly and there is a long association between mother and cub. Albatrosses breed in large colonies. In some species the birds make a bond for life and share all parental duties. Royal albatrosses do not breed until they are 8 years old. They lay a single egg a year and tend it carefully.

Years

Mayfly

Wolf

Monkey

Stork

Bear

Royal a[lbatross]

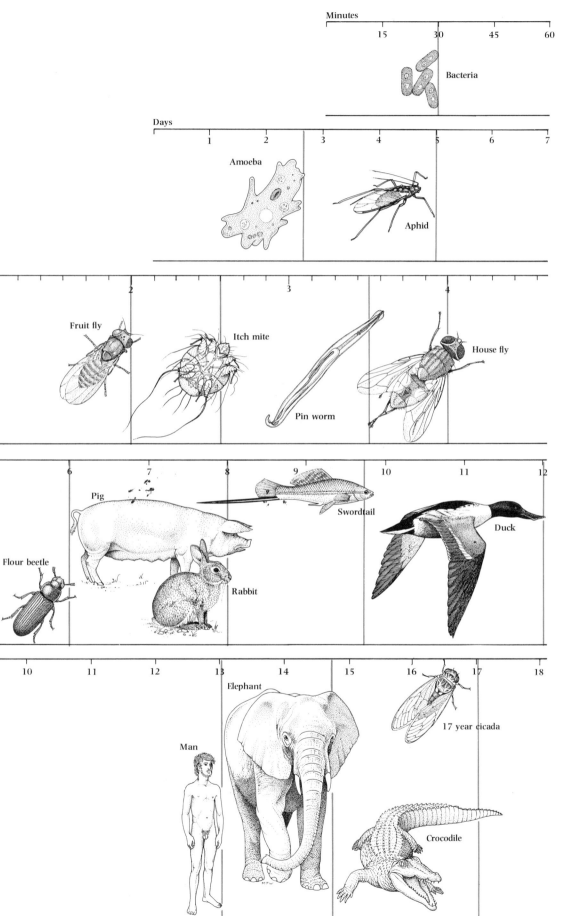

Minutes

15   30   45   60

Bacteria

Days

1   2   3   4   5   6   7

Amoeba

Aphid

2   3   4

Fruit fly

Itch mite

Pin worm

House fly

6   7   8   9   10   11   12

Flour beetle

Pig

Swordtail

Duck

Rabbit

10   11   12   13   14   15   16   17   18

Elephant

Man

17 year cicada

Crocodile

Bacteria reproduce by division. In good conditions, they divide every 20 minutes. In 24 hours a cell can give rise to some 4,000 million, million, million offspring.

Amoebas reproduce by simply splitting in half. In the summer when plant food is plentiful, female aphids reproduce from unfertilized eggs. In the autumn males are produced and fertilize eggs which survive the winter.

Insects such as the fruit fly and the house fly are mature in a few weeks. The itch mite, scarcely visible to the naked eye, also reproduces rapidly. The pin worm is a parasite of the human intestine. The female is mature in 3 to 4 weeks and migrates to the anus to lay eggs.

Pigs and rabbits are both about 8 months old when they become sexually mature. Rabbits born early in the year are capable of breeding in the same year. Ducks are mature at about a year old. They moult the first winter plumage and develop nuptial plumage in the spring for the first breeding season.

Elephants are long-lived mammals, slow to mature. They live in family groups and calves feed from any nursing mother. Nile crocodile males are over 9 ft (2.7 m) long and females 8 ft (2.4 m) before they are sexually mature. 17 year cicadas live underground at the immature nymph stage for 17 years. In a particular area on just a couple of nights, the nymphs emerge and transform into mature adults. They mate, lay their eggs and die.

Animal courtship is essentially similar to human courtship. Both animals and humans like to formalize the mating transaction with some form of ritual, whether it be a display or a wedding ceremony. The essential components of courtship are: attraction, pursuit and a period of time before mating occurs. For many animals the relationship, or pair bond, may continue while the young are brought up, or even for life.

That this elaborate pattern of behaviour should be favoured by so many different kinds of animal suggests that it has great importance for their lives and future survival. In fact, courtship has several valuable functions, its primary one being the attraction of a mate. A secondary function is to make sure that the mate is of the right species. In animals which reproduce sexually, the male and female must be in close contact so that their sperm and egg cells unite. For internal fertilization the animals must get close enough to copulate, but even for external fertilization proximity and synchronization are necessary. Female salmon, for example, lay thousands of eggs on a river bed; the male must quickly cover them with his sperm. A male stickleback builds a tubular nest of algae in which the female lays her eggs; the male must swim rapidly through after her to fertilize them.

Most animals have evolved some kind of communication system to enable them to find a mate. There are at least three different methods of signalling—visual, acoustic and chemical. The signals that an animal uses depend mainly upon its sensory equipment, which, in turn depends upon its habitat. Visual signalling is relied upon by animals which live in open habitats where they can easily be seen. Brightly coloured tropical fish, for instance, are found in clear, shallow water, and many beautifully plumaged birds such as ducks and grebes frequent wide open spaces. Woodland inhabitants use high, conspicuous perches from which to display or perform elaborate acrobatics. Some animals, such as the sage grouse and some antelopes, prefer high, open display spaces where their struttings and posturings can be seen from far away. Visual signals have the advantage of being long lasting. Once a special feather or colour is built into an animal, it can be revealed or shown off almost continuously. However, the range over which the signal is effective may be severely limited, particularly in dense habitats.

Nocturnal animals and those that live in dense, dark habitats, tend to favour sound signals. But exceptions, such as fireflies which flash through clear night skies, and some species of bioluminescent fish, can overcome the problem of visual signalling at night or in the darkness of deep sea. Since sounds carry for enormous distances underwater, whales and dolphins squeak and grunt and sing long, complicated songs to contact each other. It has been said that whales can communicate with each other over hundreds of miles. Similarly, the howls of monkeys and gibbons can be heard long before any visual contact can be made in tropical rain forests, and the calls of green tree frogs can penetrate the dense vegetation at night more readily than any visual sign.

Chemical signalling is particularly important to invertebrates in their search for a mate; it may well be the oldest method in evolution. Primitive life forms, perhaps without equipment to send and receive visual and acoustic signals, might have had acute sensitivity to specific chemicals from their own species. Female butterflies and moths, for example, secrete volatile chemicals called pheromones that can be carried miles on the wind to attract males. Mammals are also sensitive to smell; male dogs are drawn from a wide area to the scent of a bitch in heat.

Whatever method of signalling used, it must attract an individual of the right species, which can prove difficult in an area containing several closely related species. When animals get confused and mismating takes place, they can produce infertile hybrids, which means that no genes will be passed on to future generations. Even if hybrids are fertile, their fitness is lower than that of either parent. Consequently, there is tremendous pressure to avoid hybridization and each species has developed its own characteristic signals that cannot be confused with any others. Each firefly species will flash with its own particular rhythm, and each species of finch will have its own specific colour pattern. Many species of warblers and tree frogs look alike, but each species develops its distinctive courtship song, and females will respond only to the song of their own males.

A process known as ritualization has evolved in which certain activities have been modified into stereotyped and unmistakable signals. It relies upon the availability of various behaviour patterns that can be turned into these special signals. The activity may be as simple as a lizard nodding its head or a bird raising its wings before flight. Or ritualized behaviour may arise under conditions of conflict and stress such as a territorial battle. An animal may beat the ground rather than his opponent, or may try to build a nest. Such activities are thought to take place because the animal is torn between attacking and escaping, but instead performs some irrelevant action like preening its feathers. The stress involved may have physiological effects such as an increase in breathing. In fact, many signals associated with inflating special sacs, as in tree frogs or frigate birds, may have originated this way.

Regardless of the origin of the particular behaviour, ritualization guarantees that it will occur in only one form. Many signals are so constant that they can be measured to within a fraction of an inch and timed to within a hundredth of a second. Thus there can be no confusion with a similar signal of another species. Fiddler crabs use a claw for signalling which has become enlarged and coloured so as to make it more conspicuous. Similarly, many birds' feathers develop specific shapes and bright colours to enhance their signal value.

Courtship may be a protracted affair involving the exchange of a series of signals over a period of time. In the great crested grebe, for example, courtship may fulfil several important functions. The timing of breeding is crucial to make certain that the young are hatched when conditions are good and there is sufficient food available. Male and female must agree about the timing and also be physically ready to produce large numbers of sperm and eggs together. Complicated series of displays may help the birds to stimulate each other and ensure synchronization of breeding activity, as well as to signal that the next step in the long chain can proceed.

Long periods of display allow the female to assess the fitness and suitability of the male admirer. She is effectively delaying the moment of final choice and consummation until she is sure she has made the best choice. For the female this is critical, for once she has made her decision she has committed herself not only to the male but also to a long period of rearing the young. She may have only one chance to pass on her genes, whereas the male, once he has fertilized her, usually has the opportunity to mate several more times with other females. In this context, courtship becomes less of a cooperative venture and more of a battle of the sexes. The female is making a vital choice, perhaps needing the male to stay and protect her and help her bring up the young. The male is trying to convince her of his ability to do so. But, having achieved his objective, and knowing she can feed the young, he is under strong evolutionary pressure to leave her and fertilize more females. In mammals, the female is supplied with mammary glands to feed the young, and thus for some time is the only parent necessary. It is not surprising, therefore, that most mammals are polygamous. In most bird species, however, both parents are needed to feed their fast-growing young and so monogamy is more typical. Probably the only limit to the size of a male mammal's harem is the number of females he can successfully defend against rival males.

In polygamous species, courtship is an aggressive time. The females are herded into a group and jealously guarded by a large male who must frequently fight off attempts by other males to poach them. Red deer stags, elephant seal bulls and male lions engage in protracted struggles rather than delicate courtship to attain a female harem. There is great competition among the males, and only those big enough and strong enough to defend a territory or a group finally achieve success and breed. This rivalry has led to sexual differences—males get bigger, stronger and more attractive to females. When choice is involved, it is in the female's interest to select the most successful male so that her genes will be carried on. When female choice is predominant, sexual selection is operative. In some cases, however, the reason a female chooses a male with the brightest colours or biggest tail, as in the case of the peacock, is not obvious. Perhaps females select a male with a large tail because he has proved he can survive in spite of such a handicap. In most cases, however, sexual selection must act in balance with natural selection to produce males which are both fit and attractive to females.

COURTSHIP

With bright plumage, strutting displays and strident songs male animals try to prove their fitness to mate.

Visual attractions

In their attempts to attract and excite a suitable mate, many animals perform particular movements as well as display their elaborately coloured or shaped bodies. Known as visual signals, these stimuli may be as simple as a nod of the head or as complicated as a courtship dance.

A system of visual signalling depends not only upon an animal's ability to transmit light signals, but also upon its ability to receive them. Most animals are able to perceive only a small proportion of the total spectrum of radiant energy as light, although some insects such as the honeybee are capable of seeing into the ultraviolet. To a certain extent, the type of signal an animal can send is determined by its habitat. Since the majority of species are active in the daytime and live where light can penetrate, they use reflected light to advertise their presence to potential mates. But nocturnal animals, such as fireflies and their larvae (glowworms) and those which live where sunlight cannot pierce, such as the fish and squid of the ocean depths, must produce their own light biochemically (bioluminescence).

Simple repetition of a specific activity is a commonly used visual signal. During courtship, fireflies produce a bright but intermittent light, and many birds bow or nod their heads or raise and lower their wings rhythmically. Each species, however, has to produce its own characteristic pattern for an individual of the same species to respond and be attracted for pairing.

The striking colours and patterns of animals are often courtship signals. In most animal species it is the male that is gaudily attired and the female that is more soberly dressed.

Shape also has an important part to play in visual signalling. The plumage of birds, for instance, is shaped to show off particular colours or further enhance an overall design. Often, movement, colour and shape combine to produce an extremely elaborate composite signal as in a courtship display or dance. The male peacock opens and shakes his enormous shimmering tail of iridescent feathers as he parades before the female. And the many species of birds of paradise exhibit their exotic feathers to potential mates in the most intricate ritual movements, some even hanging upside down from branches.

Fireflies are nocturnal insects. The flashing light they can produce is used as a courtship signal. The males make a stereotyped pattern of flashes; the female recognizes the flash pattern of her own species and answers by her own light signals. Fireflies of the genus *Photinus* have light organs containing a substance called luciferin which, together with 2 other chemicals and oxygen, produces visible light. A nerve impulse activates this biochemical pathway to trigger the flash of light. The organs are positioned on the underside of the insect's abdomen.

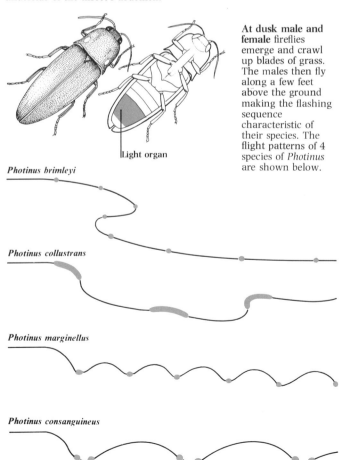

Light organ

At dusk male and female fireflies emerge and crawl up blades of grass. The males then fly along a few feet above the ground making the flashing sequence characteristic of their species. The flight patterns of 4 species of *Photinus* are shown below.

Photinus brimleyi

Photinus collustrans

Photinus marginellus

Photinus consanguineus

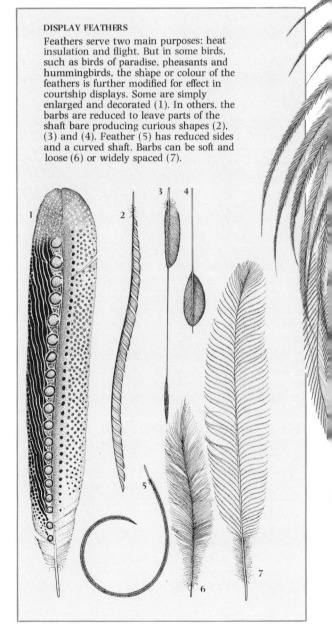

DISPLAY FEATHERS

Feathers serve two main purposes: heat insulation and flight. But in some birds, such as birds of paradise, pheasants and hummingbirds, the shape or colour of the feathers is further modified for effect in courtship displays. Some are simply enlarged and decorated (1). In others, the barbs are reduced to leave parts of the shaft bare producing curious shapes (2), (3) and (4). Feather (5) has reduced sides and a curved shaft. Barbs can be soft and loose (6) or widely spaced (7).

One of the most splendidly adorned of all birds is the Count Raggi bird of paradise, *Paradisaea raggiana* of south New Guinea. The head and neck of the male bird are yellow with some areas of glossy green or black. At his sides are startling red plumes up to 20 in (51 cm) long. Thin elongated feathers or tail wires curve out from beneath the perching bird. The female is much plainer with dull plumage.

In display the bird bends forward and moves his wings right up so that they touch over his back to make a clapping noise. He also extends and lifts the long red flank plumes. The whole performance is designed to show off to full effect the striking colours and shapes of the display feathers.

35

Competitive displays

Living space and food are continually fought for, but it is the outcome of animals' rivalry for a mate that will ultimately determine their biological success. For the only way they can ensure the survival of their own genes is to breed and leave behind as many offspring as possible.

Usually a male has to compete with other males of his own species for a mate, and this rivalry often takes place at the time of courtship displays. In most species, the male tries to persuade the female that he would make the best partner by showing her that he is capable of finding and defending a place for them to live in, as well as providing food. He will act aggressively toward other males, revealing any weapons he has, such as horns, teeth or claws. He will also attempt to make himself look as big and strong as he can through posturing and strutting. Real trials of strength in battle, or merely threatening gestures in a display, give the female an opportunity to assess the male. She must be highly selective, for her choice determines whether her offspring and genes will survive in future generations. Thus courtship involves both competition between males and the important element of female choice.

Competition also exists between different species, particularly closely related ones, for the same environmental resources such as food or nesting sites. Generally, evolution makes sure that direct competition is avoided and that each species exploits the environment in a slightly different way. However, there is sometimes an overlap, especially when closely related species occur in the same area. It is possible, for example, to find several species of crab on the same shore, or several species of lizards living in the same area. Under these circumstances, there is a danger that mating will take place with the wrong species, mainly because closely related ones look so similar. When this does happen it does not take long before the penalties of mistaken mating become apparent: the hybrid offspring tend to be a disadvantageous mixture of both species; even if they manage to survive and eventually obtain a mate, breeding will probably be unsuccessful as hybrids themselves are usually infertile.

To avoid any confusion about the species of a signaller during courtship, each species has perfected its own distinctive version of a signal. In the case of crabs using visual stimuli, for instance, no two species wave their claws in precisely the same way, just as in semaphore each letter is clearly represented by only one position of the flags. The head nodding of *Sceloporus* lizards is an excellent example of a species-specific signal in action. Female lizards are attracted only by the particular nodding pattern of their own species and avoid mating with any other related males. This has been demonstrated by an electronically controlled model that can be made to nod to the pattern of any species. Only when the pattern conforms to that of her own species will a female lizard move toward the nodding model and prepare to engage in courtship and mating.

During courtship a male *Sceloporus* lizard approaches a female and performs a head-nodding display. He moves his head quickly up and down in a series of nods. These nods last a second or two and the head moves less than an inch up and down. Even though the display is very fast and the movements tiny, the signal is enough to interest a female of the right species who will then start to respond to the male.

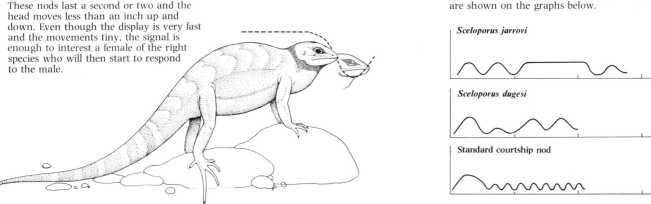

Each lizard species has its own version of the nodding display; the females only respond to the nod of the male of their species. Some different nodding sequences are shown on the graphs below.

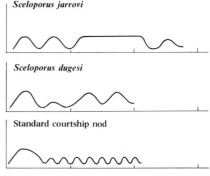

Sceloporus jarrovi

Sceloporus dugesi

Standard courtship nod

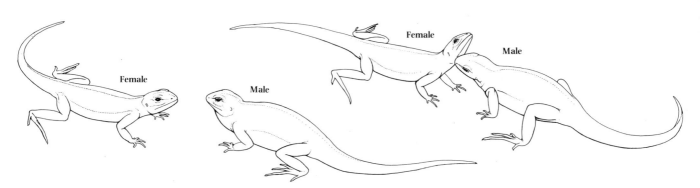

If the correct nodding pattern is displayed to a receptive female, she will swing her tail from side to side in a series of arcs, each swing lasting about a second. This is her signal for the male to continue his courtship. While she remains still he begins to nudge her and bite her neck. If she attempts to move away he repeats his nodding display until she is still. When it is clear that he has been accepted, the male may bite his mate's neck and keep this hold while he attempts to mate her. He twists his tail under hers so that their cloacae are pressed together. Copulation lasts only a few seconds.

Many claw-waving displays exist; the pattern of each species of fiddler crab is slightly different. Some crabs such as *Uca lactea* have a complex lateral wave, rather like a beckoning movement. Others such as *Uca signata* and *Uca zamboangana* make simple vertical waves. The subtle but constant differences are sufficient to allow the females to recognize a mate.

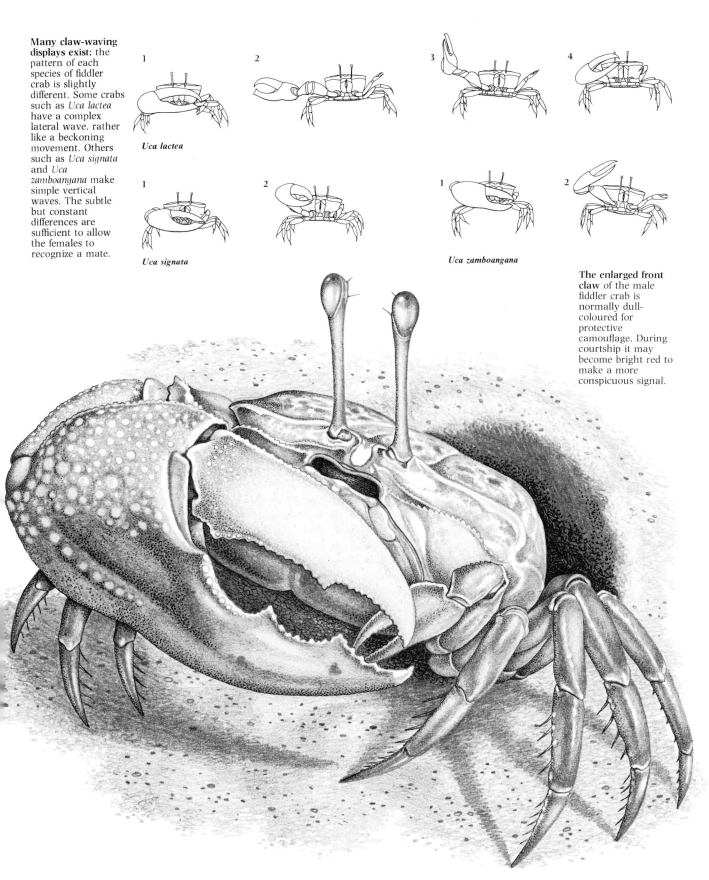

1   2   3   4

Uca lactea

1   2

Uca signata

1   2

Uca zamboangana

The enlarged front claw of the male fiddler crab is normally dull-coloured for protective camouflage. During courtship it may become bright red to make a more conspicuous signal.

Fiddler crabs of the genus *Uca* perform a claw-waving courtship display. One front claw (cheliped) of the male crab is much larger than the other and brightly coloured. The male crab stands outside its burrow on the shore and waves this huge, vivid claw to attract a female's attention. The crab must use its other normal cheliped for practical tasks such as feeding, as the signalling claw is so large that it is useless for anything else—an example of how a part of the body can be modified for a specific role. The fiddler crab, above, *Uca marionis*, is found in North Australia, the Philippines and Fiji.

How a display evolves

The bizarre movements or signals of birds performing a courtship display often have their origins in simple everyday behaviour such as preening or flying. The movements appear odd only because they have been taken out of context and incorporated into other situations, or because they have been exaggerated or even combined with quite different movements in a display.

To ensure that there is no confusion during courtship, the specific signals that animals send to other members of their species are always performed in the same pattern; the process by which they have become fixed is known as ritualization. A bird waving its wings in a normal, unritualized way suggests a bird about to fly. But in a ritualized form, this movement could change in a number of ways: it might happen slowly or quite fast; it might become exaggerated and extend above the head or it might include the display of particular coloured feathers. Similarly, if a human was trying to attract attention by waving his arm, he might extend his arm fully and make slow extravagant movements, or he might add a handkerchief for colour and effect.

Conflict behaviour, which is exhibited when an animal is torn between opposing drives or motivations, has also provided a rich source for displays. A male trying to court a female may be motivated aggressively—in some ways, she is an intruder into his territory—as well as sexually because he wants to mate with her. The female may likewise be in conflict, motivated sexually like the male but also by fear that he may attack her. Under such circumstances, an individual may be unable to make the appropriate response and sometimes performs some irrelevant activity such as preening or feeding. Either movement occurs out of its normal context and in situations of high stress.

The inciting displays of female ducks are among the most beautiful examples of the difference between ritualized and unritualized behaviour patterns. After pairing, both male and female help to defend their territory, and the female often warns and encourages the larger male to attack a rival by pointing repeatedly toward it with her head and neck. In the unritualized form of the action, the female shelduck, for instance, points toward the rival with her head no matter which way her body is pointing, and so the display has a variety of forms. In the female mallard, however, it has become highly ritualized and only occurs as a stereotyped pointing over the left shoulder, irrespective of the direction from which the rival approaches.

The movements a bird makes when it is about to launch itself into the air are the basis of courtship rituals known as flight intention displays. The heron, below, is actually about to take off and is bending its legs and lifting its wings in preparation for flight. The cormorant, however, is performing a flight intention display; its wings are raised and the head is tilted back in an exaggerated posture.

Elaborate ceremonial dances are performed by the crowned cranes, *Balearica pavonina*, which are widely distributed in Africa. The 20 in (51 cm) wings are outstretched to display the feathers to full advantage and the birds strut about bowing their heads and jumping into the air. This crane takes its common name from the tuft of stiff, yellow feathers set behind a black cap of shorter feathers on its head.

Heron

Cormorant

Preening—real or mock—often forms part of the courtship of ducks. In the display of the shelduck, *Tadorna tadorna*, the male vigorously preens his feathers.

The garganey, *Anas querquedula*, shampreens during courtship, concentrating on the outside of the wing to make the bright blue feathers at its base conspicuous.

The **mallard,** *Anas platyrhynchos,* makes as if to preen an area in front of a bright patch on the wing, raising its wing to reveal the special colours.

The **mandarin,** *Aix galericulata,* has one particularly large and vivid feather on the wing. Instead of actually preening, the duck points toward the special feather. One of the most elaborately plumaged of all ducks, the mandarin makes other courtship movements based on normal behaviour such as drinking.

Signalling by sound

Attracting a mate by visual signals such as displaying feathers or waving a claw, is not always feasible. Animals that live in dense vegetation or those that are nocturnal and active only in darkness, have mostly abandoned attempts at visual signalling and use sound signals to find a partner.

Any movement in a medium such as air causes disturbance or sound waves; these are changes in pressure which radiate out like ripples from a stone thrown in a pond. Animals, from insects to man, have mechanical devices for producing sound which control the frequency and size of the sound waves they create. The larger the disturbance made, the louder the sound waves will be. The faster the disturbances, the closer together the individual sound waves are and the higher the frequency or pitch of the sound produced.

The perception of sound quality or tone depends on the type of sound which the sense organs—ears—can detect.

Most have a membrane which vibrates in response to the pressure changes caused by sound waves and transmits these to sensory cells which analyze the information. Some types of ears are sensitive to a wide range of frequencies; others are finely tuned to respond only to the range of frequencies their species uses when signalling. Man hears best in the range 1–20 kilohertz; many moths hear from 20–100 kilohertz. Moths need to hear in this range to detect the high frequency sounds made by the bats which prey on them. The effectiveness of the communication depends upon both the transmitter and the receiver.

A wide range of sound-producing devices exists in the insect world. The mosquito uses wing vibrations, other insects like grasshoppers, rasp a filelike part of the leg against a ridge on the wing; the cicada has a structure on its abdomen containing a disk which is buckled by muscular movements to make click sounds. These are simple

noises, but each species of, say, cricket has developed its own characteristic sound to attract only mates of its own species—vital when physical differences may be slight. The timing and spacing of insects' chirps and trills vary, and experiments with crickets show that females ignore the songs of other species.

Among vertebrate creatures, amphibians, birds and mammals make use of vocal communication. Bird songs are varied and complex and play an essential role in courtship but many mammals have a relatively small repertoire of grunts, howls or barks. A notable exception is the humpback whale, *Megaptera novaengliae*, which needs to use sound signals to find and keep in touch with its mate in its deep dark ocean home. The humpback has a huge range of separate sounds used to compose songs which last from 7 to 30 minutes. These songs are probably the most complex sound displays known in animals.

A rattling or rasping noise is made by courting grasshoppers. The sound mechanism has 2 parts: a file of raised pegs on the inner surface of the leg and a scraper or hard ridge on the wing. As the hind legs move up and down, the file scrapes against the ridge to produce sound.

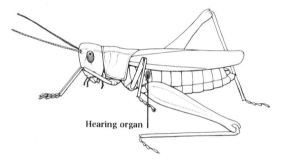

Hearing organ

Femur

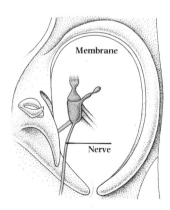

Membrane

Nerve

The hearing organ of the grasshopper on each side of the first section of the abdomen is an oval membrane surrounded by a thick rim of skin. The membrane vibrates in response to sound waves; these waves are eventually converted to nerve impulses in a special nerve ganglion and transmitted to the nervous system.

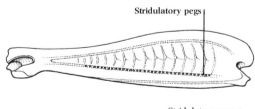

Stridulatory pegs

Stridulatory pegs

The file, in this species of grasshopper on the femur section of the leg, is a row of tiny knobs or pegs. The size, number and density of these stridulatory pegs vary between different species but on a file 0·12 to 0·20 in (3 to 5 mm) long, there can be 80 to 120 pegs and on a file 0·16 to 0·28 in (4 to 7 mm) long, 300 to 450.

Species of male grasshoppers all have their own characteristic song pattern so that the female can recognize and mate with the right species. Variations in the organization and timing of songs are brought about by different firing patterns in the central nervous system, as well as varying arrangements of the stridulatory pegs. The hearing organ can discern sounds separated by only a few hundredths of a second.

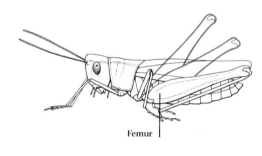

Gomphocerus sibiricus

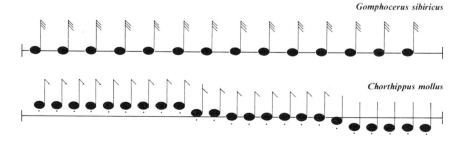

Chorthippus mollus

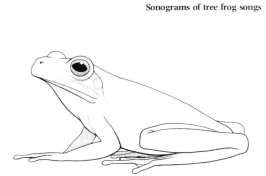

Sonograms of tree frog songs

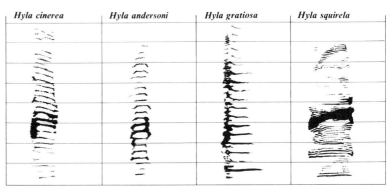

| Hyla cinerea | Hyla andersoni | Hyla gratiosa | Hyla squirela |

Amphibians, such as frogs and toads, were the first vertebrates to have vocal cords contained inside a larynx. Air from the lungs is forced over the vocal cords causing them to vibrate and create sound waves.

A vocal sac beneath the chin which inflates and acts as a resonating chamber to amplify the voice is a feature common to many frogs and toads. Before calling, the lungs fill beyond their normal capacity. Trunk muscles contract and force air into

the sac and over the vocal cords. After the call, muscles in the sac contract and the air goes back to the lungs. Mating calls are short, harsh barks or croaks. Each species has its own characteristic version of these calls.

This tree frog, *Hyla citroba,* is about 2 in (5 cm) long. Its vocal sac is fully inflated.

Why birds sing

Perhaps the most beautiful of all animal sound signals are the songs of birds. A song is a long, complex sound signal given only by male birds and particularly linked with mating. The main reason why birds sing at all is to attract and court a mate, although songs also contain other information such as territorial signals. Simpler calls are used for other aspects of vocal communication such as giving alarm calls to warn others of the approach of a predatory bird.

A single male may sing all day but once he has attracted a female he sings much less—evidence that sexual attraction is the main purpose of song. Some birds, such as the sedge warbler, stop singing suddenly and completely after pairing and do not sing again until the following year.

If the females are choosing males by their songs rather than by their plumage or display, there is bound to be tremendous competition among males to produce more and more elaborate versions. An individual sedge warbler male may have over 50 different sounds, syllables, in his repertoire. When he sings a song it may last over a minute and consist of several hundred syllables. He uses about seven different types of syllable on average in each song, repeating, alternating and mixing them in a particular way. He starts afresh for the next song by selecting a few other syllables from his repertoire. Each time he composes a completely new song, producing thousands of variations on the basic theme. Songs include clicks, buzzes, whistles and chirps.

Other birds, such as peacocks and birds of paradise, rely on their ornate plumage to attract mates. In some ways the song of a plain little bird like a sedge warbler is the sound equivalent of a peacock's tail or an exotic feather display—each is the ultimate signal of its own kind.

Elaborate songs are also important in species recognition. Many species of small birds such as marsh and reed warblers look alike and live in the same marshland areas, but each evolves its own type of song so that females know when they have found the right male. Recognition in birds depends on various cues contained in songs. In the songs of European robins a high note must always alternate with a low note, and artificial songs constructed on this basis and played from a tape recorder attract other robins. Similar experiments with the songs of American buntings show that the spacing or ordering of the notes in their songs is important to species recognition.

Just as there are many dialects of the same language in human speech, some species of birds have regional variations or dialects in their songs. White-crowned sparrows of the San Francisco Bay area of the United States have slightly different versions of the same song according to where they live. Experts can accurately place the origin of an individual bird by the dialect of its song.

The function of dialects in bird song is not fully understood, but it may be that the best place for a female to breed is in her own home locality, for which she is particularly well adapted. What better way for the female of a highly mobile and vocal species to recognize a suitable mate than by his own distinctive home dialect?

50 different syllables may be in the repertoire of a male sedge warbler. Sonograms, below, give a picture of the sound by plotting frequency in kilohertz against time in seconds. A vertical line is a wide frequency spectrum produced in a short time and sounds like a click. Several clicks together sound like a buzzing noise or rattle as in (1). A single line moving horizontally is an ascending or descending whistle as in (2). The sounds in the third group are like chirps and the longer, wider frequency sounds in (4) are barking noises.

The sedge warbler, *Acrocephalus schoenobaenus,* lives and breeds in marshy areas throughout most of Europe. A small brown bird about 3½ in (9 cm) long, it makes up for its lack of visual impact in 2 ways. First, the male has one of the longest, and most elaborate of all bird songs which he uses solely for the attraction of the female. Second, he enhances the long-distance transmission of the song by selecting a high song post such as the top of the tree or by performing special song flights as above. He gains height first and then descends in a slow spiral singing all the while.

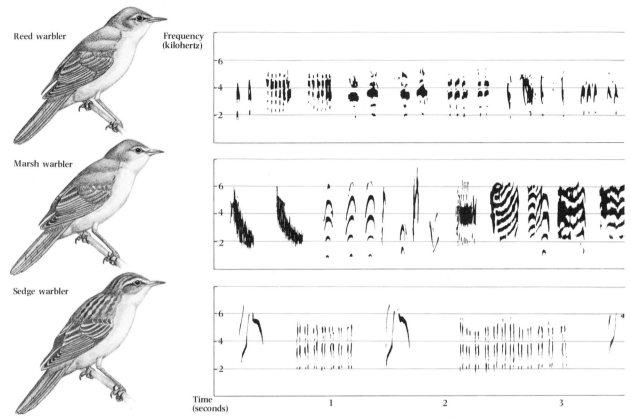

Reed, marsh and sedge warblers, all *Acrocephalus* species, are found in the same marshland areas. To recognize their own species correctly they have developed quite different songs. All have their own specific syllables as shown by the shapes on the sonograms above. The overall frequency ranges also differ: the reed warbler song is the lowest, the sedge the highest and the marsh has the widest range. The reed and marsh warbler favour doublets and triplets, whereas the sedge warbler uses recurring alternating phrases. All these warblers are small birds about 3½ in (9 cm) long.

Each local population of white-crowned sparrows, *Zonotrichia leucophrys*, around the San Francisco Bay area has its own particular dialect of the same basic song. The sonograms above show 2 songs from 3 populations of birds. The birds in the Berkeley area make 3 or more ascending whistles at the start of a song, followed by a trill. The Marin birds have a longer whistle followed by a trill, and the Sunset Beach birds 2 ascending whistles followed by a trill. The birds are about 6¾ in (17 cm) long. Their plumage is light and dark grey with streaks of brown, black and white.

43

Insect aphrodisiacs

As if by some immensely powerful magnetism, a male moth can detect a female of his own species from more than a mile away. He cannot see or hear his potential mate at that distance but, by use of his feathery antennae, can smell her. Flying into the wind, he homes in up the odour gradient until he is near enough for courtship and copulation.

The odorous substances produced by female moths, and by members of all insect families as an essential ingredient of courtship behaviour, are chemicals called pheromones. Their precise action is not known but the responses they produce are so predictable that pheromone-baited traps are widely used in man's war against the insects which damage his crops and cause disease.

In insect courtship, pheromones do more than merely bring the sexes together. Some seem to act as aphrodisiacs, goading male or female to start an appropriate repertoire of courtship behaviour. The butterflies and moths of the insect order Lepidoptera have a huge range of scent-producing organs. These may be on the legs, the central section of the body (thorax), the wings or the abdomen. The pheromones these organs release also have a calming effect on the female and help overcome her natural urge to fly off as the male approaches.

As well as encouraging courtship behaviour, pheromones act as a barrier to interbreeding between different species. If the correct pheromone is not present, courtship stops immediately.

Not all insect pheromones are exclusively recognized by the members of one species. The males of several species of pyralid moths will all, for example, respond to the sex-attractant pheromone emitted by the females of all the species in the group. Confusion is avoided in the wild because each species is active at a different time of night. In other closely related species of moths the only variations are in the concentration of pheromones, but again added insurance cover is provided by differences in activity patterns.

Outside the insect world our knowledge of invertebrate pheromones is scanty, but earthworms, and crustaceans such as lobsters and crayfish, certainly rely on special chemicals to bring the sexes together. In the history of animal life on earth chemical attraction seems to be almost as old as sexual reproduction itself. The odorous substances used by today's animals for finding a mate probably evolved by accident from wastes associated with the reproductive processes of each partner in the courtship contract.

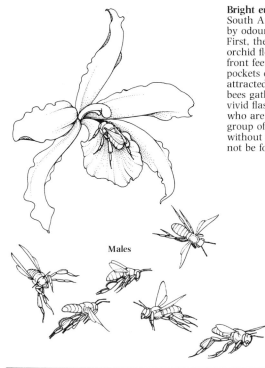

Males

Female

Bright euglossine, or golden bees, from South America have taken the part played by odour in courtship one step farther. First, they sponge up fragrance from orchid flowers with special pads on their front feet. This fragrance is transferred to pockets on the hind legs. Other males are attracted by the fragrance, and a group of bees gathers. The general excitement and vivid flashing of colour attracts females who are then mated. The colour of the group of bees attracts the females, but without the scent the mating group would not be formed.

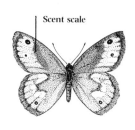

Scent scale

Once the female has alighted, the male grayling butterfly, *Eumenis semele*, walks round to face her and starts to jerk his wings. Appearing to bow low before the female, he closes his wings, trapping her antennae between special scent scales. The odour released transfixes her and he quickly mates.

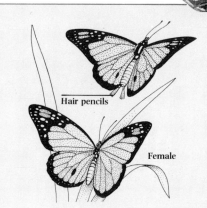

THE COURTSHIP DANCE
Visually alerted to the presence of the female queen butterfly, the male soars up to hover just a little above and in front of her. By giving out odour from his hair pencils he persuades her to alight. He copulates when he alights on the female's back between her upswept wings, and during mating he strokes the female's antennae as if to calm her. After copulation the male flies upward still coupled with the female who lies passively beneath him. They fly off to a quiet place and remain locked together for several hours.

Hair pencils

Female

The male queen butterfly, *Danaus gilippus berenice*, courts his mate with scent dispersed by a pair of hair pencils at the tip of his abdomen. He flies close to the female and liberally douses her head and antennae with the odour. As he does this he bobs up and down and not infrequently bumps into her. The presence of hair pencils is essential for successful courtship. Males that have had their hair pencils removed go through the motions of courtship but cannot persuade the female to stay on the ground long enough for copulation.

Hair pencils

Each tuft of 'hair' is composed of about 400 scales which superficially resemble hairs. These are normally pulled up inside the abdomen out of sight. Special muscles push them out and pump secretion into the hairs from the gland at their base.

The lure of scents

Most mammals depend more on smell for communication than on their eyes and ears. In common with other animals with well-developed olfactory senses, such as fish and reptiles, they use olfactory signals, pheromones, as a stimulus in courtship.

Attraction and signalling of sexual status is the first phase of courtship. In some female fish the ripe ovaries secrete small quantities of pheromone which are readily detected by males. But in the blind goby and the goldfish, the pheromone is derived from sex hormones that normally circulate in the blood. When female mammals excrete pheromone in their urine, males exhibit intense courtship behaviour. They do not perceive the odour in the nose, but in the accessory olfactory organ of Jacobson—a blind-ended pouchlike structure that lies immediately above the hard palate and opens into the mouth through a fine duct. This is usually closed but can be opened by the mammal curling back its upper lip, as if making a gesture of disgust; the opening of the duct allows the odour to travel to the sensory organ. This behaviour, known by a German word, *flehmen*, is widely seen in cats, ungulates and marsupials.

Pheromones not only attract mates but also signal the state of readiness of the female's heat. The boundary marks of solitary territorial species such as the big cats, which are visited by all holders of adjacent territories, carry information about the sexual state of the occupier. During the heat period, the female produces a strong scent from all over her body that acts as a magnet for all the males within miles.

The second phase of courtship is mating, and in big cats among others this happens shortly after the sexes have come together. But in gregarious species, which live together all the time, it is sometimes necessary for the male, who normally precedes the female into breeding condition, to induce sexual readiness in his mate. Experiments with rodents and sheep reveal that it is odorous substances in the male's urine that initiate estrus in the female. Females will respond to the smallest amount of these odours, which are derived from male sex hormones. In one mammal, the pig, the odour of the urine, saliva or even the breath of a male is sufficient to cause the female to adopt the mating posture if the estrous cycle is at its peak. If it is not, she is unaffected and walks away. Pig breeders regularly use synthetic porcine pheromone in aerosol form to enable them to detect which of their sows are ready for service.

Courtship pheromones are probably much commoner in vertebrates than is currently known. It is thought that the strong odour produced by the preen gland of the kiwi at the onset of breeding is a courtship pheromone. Whether such substances exist in human beings and if so whether they play a role in courtship is still a much debated topic.

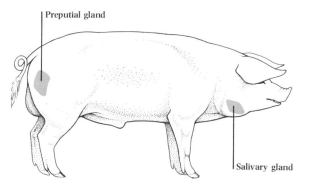

The salivary gland of a boar carries a compound derived from the male sex hormone testosterone. While courting the female, the boar salivates profusely, covering himself and much of his surroundings with the pheromone. The sow is highly sensitive to the male's odour.

Preputial gland

Salivary gland

The Essex sow, below, is in the mating posture induced by the boar's odour; she has ears erect and her back straight. The male pig pheromone is also produced in the preputial gland at the base of the penis. Some of this secretion mixes with sweat and urine and gathers in the preputial pouch. During mating some of the substance rubs off on the female, perhaps signifying that other males would be wasting their time to mate with her. To the human male the pig pheromone smells slightly of urine—not pleasant but certainly not repulsive. Women, however, find the odour overwhelmingly pungent, smelling strongly of rancid urine.

Courtship behaviour in many snakes is controlled by pheromones produced by specialized glands on the female's back. The common garter snake, *Thamnophis sirtalis*, of the United States, has many subspecies. The courting male runs its chin along the female's back. By her odour he knows whether or not she is the correct mate.

The female blind goby, *Bathygobius*, lives in a shell or crevice buried in the sea floor. At breeding time she produces an odour from her ovaries which initiates courtship in the male. Stimulated by the odour, the normally belligerent male goby dances in front of her making elegant fanning movements, and eventually leads her to the nest site. Without the pheromone, courtship cannot begin.

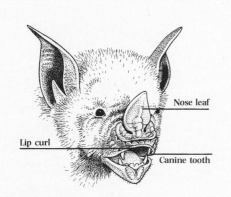

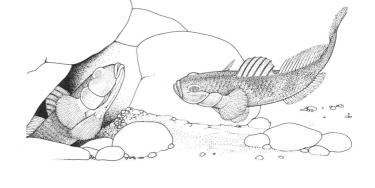

THE LEAF-NOSED BAT

Immediately after smelling the urine of a potential mate, many male mammals raise their upper lip as if in disgust. This lip-curl expression opens a duct which leads from the mouth into Jacobson's organ—a sensory organ lying just above the hard palate—which analyzes the smell. Nerves connect the organ with the accessory olfactory bulb and then with the part of the brain which controls sexual behaviour. The action is shown here for the leaf-nosed bat, *Phyllostomus*, but it occurs in cats and in most hoofed mammals.

After chasing the female, the male porcupine, *Erithizon dorsatum*, rears up on to his hind legs and directs a jet of urine on to her. This action may arise from the frustration caused by the conflict in the male animal between his sex drive and his fear of the other animal. Certainly the scent marks the female and probably dissuades other males from attempting to court her.

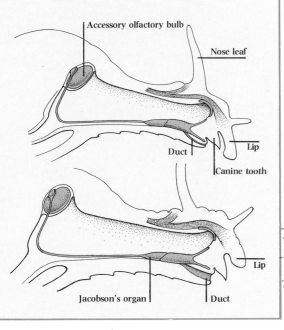

COURTSHIP
An inevitable ritual

The courtship routine of a male and a female is rather like a ballet built from *pas de deux* and pirouettes, for one link in the chain of courtship behaviour precedes the next in an orderly fashion. But while two different choreographers can take the same basic ballet movements and arrange them in different orders to the same music, the sequence of movements in an animal's courtship has no such flexibility. Each link in the chain of courtship behaviour serves as a vital stimulus to release the next, which will only appear if its predecessor has been correctly performed.

In the highly complex courtships of some animals, such as the great crested grebe, the omission, or even the premature curtailment, of one part results in the collapse of the whole sequence. Far from being a disadvantage, these conditions governing success act as a kind of insurance policy. Just as the knowledge of a secret password protects against subversive infiltration, they act to prevent members of one species accidentally interbreeding with those of another.

Because the rules governing courtship are so rigorous, animals must possess a perfect means of recognizing the links of stimuli in their own behavioural chain from all others, even those that are almost identical. Although courtship behaviour may appear complex to an observer, the key stimuli of which it is composed are in fact simple. Once each key, such as a nod of the head, is encountered in its correct form, it unlocks a releasing mechanism governed by the nervous system and allows access to the next part of the courtship ritual.

The innate fail-safe mechanism is sometimes a simple saver of time and effort. In the breeding season, male frogs, for example, cannot tell the sex of a partner by any cues received via their senses of vision, hearing or smell. Instead they respond to any nearby movement by clasping the object—whatever it is. As frogs mate communally in small ponds one male is more than likely to grab a frog of his own sex. The clasped male breaks the chain by uttering a quick burst of calls and both depart at once to find females.

When the male clasps a female no such burst of calls is given and the female adds the next link in the chain by arching her back slightly and raising her head. This stimulus releases the next key act by the male. He forms a 'basket' with his legs over her cloaca, the exit of her reproductive system, to catch and immediately fertilize the eggs that she spawns.

Advertising

Great crested grebes, *Podiceps cristatus,* are monogamous birds; the pair bond must be firmly established before they nest. The male—a large bird about 29 in (74 cm) long—advertises his presence to the female by his crests and expanded ear flaps. As he approaches her, he adopts the 'cat display' posture, hunching his neck while spreading both wings to expose their white markings. The birds wag their heads at one another, pointing their bills up and then down.

'Penguin dance'

Head wagging often ends in a ritual 'penguin dance'. Both partners dive to the bottom of the pond, grab a piece of weed and rear out of the water right in front of each other—a form of ritual feeding. Sometimes head wagging ends in the retreat of one bird. It swims off only to face its partner once again and adopt the 'cat display' posture. After repeating the sequence a few times the birds go off to build a nest.

Rearing display

The final stage of courtship, the mating ceremony, begins when the nest is ready. The male stands on a specially constructed platform of reeds near the nest and attracts the female by his head movements. She joins him, drops her head and flattens her crest. The male remains in full display with crests and ear flaps erect. They copulate amid much calling and flapping of wings from the male. The process may seem awkward for such graceful birds.

Discovery ceremony

Head wagging

Retreat ceremony

'Cat display'

Inviting display

Mating

49

The sex drive wins

Pursuit is a phase of courtship in which, having found her by visual, sound or scent signals, the male chases the female and tries to persuade her to mate. Pursuit may take a considerable time, up to an hour for tree squirrels, or be brief, a couple of minutes for most carnivores. Sometimes the chase is energetic, as in kangaroo courtship, with the male trying to grab the female by her tail. In other species, such as the antelope, it is refined and restrained. In common with other physical and behavioural attributes, the pursuit has been gradually moulded by selective pressures to fit the needs of a particular species.

The length of the pursuit is closely linked to the female's state of sexual readiness. Among chimpanzees, for instance, when a male approaches a female, she usually presents herself for mating right away. But if she is not yet at the peak of estrus and at her most receptive, she does not adopt the mating posture. The male then performs his stylized pursuit by swinging from branch to branch in front of her. He may continue on and off for several hours until she accepts him. The chase seems to help speed up the onset of the female's most receptive stage when she is ready for mating.

A mixture of adult and juvenile behaviour is often evident in the chase. To impress the female with his maturity and strength, the male acts aggressively. But at the same time, to avoid scaring her away, he also makes appeasement gestures to try and tell her she is safe. If these movements fail, however, the male resorts to a more basic trick. As the female starts to move away, he utters a few 'baby calls,' similar to those which infants make to maintain contact with their mothers. A courted female cannot resist this call and quickly returns to the male. This ploy never fails, though the male may have to repeat it several times until the female's sex drive overrides her fear of his aggression and she will allow the male near her.

The final stage of the pursuit involves a show of real aggression: animals which lack weapons, horns, tusks or antlers, wrestle with their necks; those with weapons reveal them to their partners but carefully refrain from using them. Aggression occurs at this stage because the female has to be subjugated by the male, no matter how strong her sex drive might be. For, above all, courtship is a conflict between aggression, fear and the sex drive, and although the sex drive finally wins, the other two forces are always just beneath the surface.

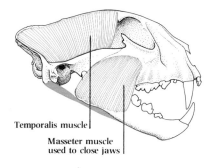

Temporalis muscle

Masseter muscle
used to close jaws

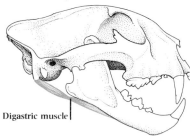

Digastric muscle

Male leopards, like all carnivores, bite the female's neck while mating. If the external masseter muscles which close the leopard's jaw were to operate as they do when he kills prey, the female could be injured or killed. Fortunately these muscles are neutralized by the digastric muscles which exert an opposing force to prevent the jaws from closing.

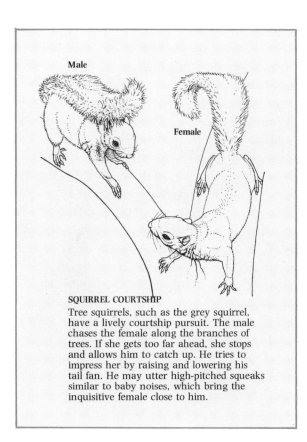

Male

Female

SQUIRREL COURTSHIP
Tree squirrels, such as the grey squirrel, have a lively courtship pursuit. The male chases the female along the branches of trees. If she gets too far ahead, she stops and allows him to catch up. He tries to impress her by raising and lowering his tail fan. He may utter high-pitched squeaks similar to baby noises, which bring the inquisitive female close to him.

The final stage of the courtship of zebras, camels, giraffes and other hornless ungulates is a violent bout of neck wrestling. The male tries to subjugate the female by sheer force but sometimes he is evenly matched by his partner. Both froth at the mouth profusely and their forequarters become flecked with spume. They can draw blood in their attempts to bite their partner's neck.

For antelopes, pursuit is a stately parade rather than a chase. In the forest-dwelling species, such as the kudu, the male walks behind the female, keeping his head low and his horns lying along his back. Only in this way can he prevent the female being frightened off by his armoury. In the gregarious grassland species, such as the Thomson's gazelle, the female is far more tolerant and the male can hold his head high when walking behind her without causing alarm.

Kudu

Thomson's gazelle

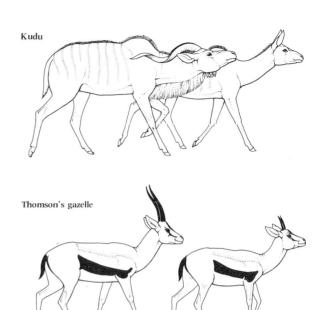

The conflict between aggression, fear and sex drive is clearly seen in the mock sparring, snarling and nuzzling approaches of courting big cats. Because physical subjugation appears to be an essential part of courtship, the male leopard, *Panthera pardus*, like all male cats, uses his teeth to grasp the female's neck at the moment of copulation. Special muscles work in opposition to the muscles that normally close the jaws.

51

Harems and couples

Harems are one form of mating system. The male fights for a band of females with which he, and only he, will mate. These males express their dominance by the size of their harem. As a rule, passing females are more attracted to large harems than to small ones. This is the polygamous system.

Monogamous animals are strictly territorial. The males compete over territories for mating, and defend them against any rivals that try to depose them. Beavers, gibbons, some species of deer mice, wolves, coyotes and 91 per cent of birds are among the creatures believed to make a monogamous bond for one or more breeding season.

The social behaviour of the male in both types of mating system serves to identify the strongest and fittest members, and only they are allowed to breed. Most male finery, whether it is visual—antlers, horns and manes; acoustic—the songs of male birds; olfactory—scent glands in male mammals—has developed specifically to enhance the animal's competitiveness.

To a large extent, the type of mating system depends upon the environment. For a polygamous or harem-defending system to evolve, a male must be able to protect his harem against the attention of other males and defend the natural resources that his harem requires. He must be able to do so economically in terms of time and energy.

In polygamous species only a small percentage of the males actually breeds and because of the competition their adornment has become highly specialized. The males of elephants, fur seals and red deer, for example, have gradually assumed a distinctly different appearance from the females.

Monogamous males, on the other hand, only have to defend one mate, or sufficient resources for one mate, and consequently, in many monogamous species there is little apparent difference between male and female. The differences that do exist appear to be for mate-attraction purposes.

Monogamy, however, must not be looked upon as a less highly evolved mating system than polygamy, for there is another important aspect of evolutionary selection that operates to support it. In many monogamous species, and in those in which one territorial male has a couple of females under his control, both parents must assist in the rearing of the family. In contrast, harem masters may father several dozen youngsters each breeding season and take no part in their rearing. If the female can protect and feed them on her own, there is no evolutionary pressure on the mate to help.

Male roe deer, *Capreolus capreolus*, are territorial animals and mate at the height of the summer social activity. The fertilized egg lies dormant in the doe's genital tract for some months before normal development starts. This ensures that fawns are born in the early spring when the best food is available.

At the start of the breeding season male sea lions come ashore on the breeding beaches and display to one another. A week or so later the females, pregnant from the previous year, come ashore and are rounded up into harems by the bulls. The females give birth almost immediately and are then mated by their harem master. When the pups are weaned 5 or 6 weeks after birth, the harems disperse. The fertilized eggs lie dormant in the newly pregnant females until development starts some months later. This ensures that the pups will be born during the next season's spell on dry land.

The ideal time for animals to mate is when social activity is at its height, and the ideal time to give birth is when food supplies are good and the weather suitable for the baby. But for some animals the length of their pregnancy means that if they mate at their ideal time the young will be born too early in unsuitable conditions. The phenomenon of delayed implantation remedies this and allows both events to take place at the best time. The fertilized egg is held in the reproductive tract of the female in a state of suspended development. After lying dormant for a few months, normal development of the egg begins and the young are born at exactly the right time. Most seals and sea lions seem to have delayed implantation, as do many small carnivores such as otters and badgers. The hormonal control of the process is still not understood.

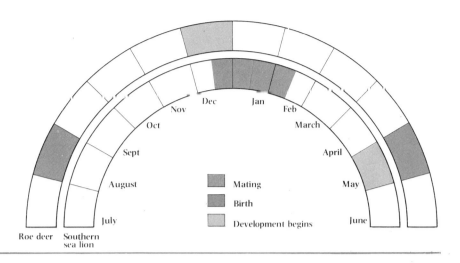

Nov Dec Jan Feb
Oct   March
Sept   April
August   May
July   June

Mating
Birth
Development begins

Roe deer   Southern
           sea lion

The Southern sea lion, *Otaria byronia*, lives off both the Atlantic and Pacific coasts of South America. Adult males are about 8 ft (2.4 m) long and weigh up to 700 lb (318 kg). Females are 6 ft (1.8 m) long and weigh about 300 lb (136 kg).

Innate in the make-up of all animals is a desire to keep their distance from others of their own kind. Even the most gregarious of creatures, such as schooling fish or herd-dwelling antelopes, respect the individual distance—the small, moving space around the body of an animal which is its exclusive property. While animals live peaceably together as long as they keep outside this space, infringement of it brings about a rapid change in behaviour which often involves the release of aggression. In vertebrate animals aggression is a common code of behaviour. Unlike the killing conduct of carnivores, with which it shares many common characteristics, this type of aggression serves to establish the social hierarchy in gregarious species and to set up and maintain territories in solitary ones.

Although individual distance is vital to the well-being of animals, many biological activities, from the act of mating to the rearing of young, demand the closest of physical contact. Because the tasks of keeping the social hierarchy intact and of defending territory are so important to day-to-day survival, animals retain their aggressive instinct at all times. They also have the means of appeasing and defusing that aggression when the normal rules of conduct must be broken to ensure the continuation of the species.

If reproduction is to be possible, a bond must be established between a male and a female for a period of time that depends on the species and its particular way of life. At the minimum, natural aggressiveness has to be appeased for just as long as it takes the male to copulate with the female. For the praying mantis, however, no pacification of the female's hungry eye is possible, and the male is invariably decapitated. To overcome this apparent setback, the removal of the head actually releases copulatory movements in the headless body and fertilization takes place posthumously.

During mating most male spiders also dice with death and resort to a series of tricks in an attempt to placate the female. Some employ subtle techniques, such as gentle stroking, while others use rather crude tactics such as tying down the female with a rope of silk. The males of predatory empid flies overcome the threat to their lives during mating by creating diversions. Each male presents his bride with a tasty morsel just before mounting, in the hope that she will be so concerned about the titbit that she will not realize what is happening to her. Surprisingly, perhaps, such tricks do work. In solitary cats and other carnivores, such as the spotted hyena, the appeasement of aggression during mating is short-lived indeed. When she is in heat and about to release viable eggs from her ovaries, the female puts out a male-attracting scent but is still rather nervous. The male approaches with his eyes diverted and frequently turns his head away. This avoidance of her gaze is enough to break down the female's barrier of reserve, and she lets the male mount her. Then he is gone and the female is left on her own to bring up the litter of youngsters.

For the many animals that pair for long periods, bonding behaviour must be strong enough to provide a favourable environment for both mating and the rearing of the young. In many mammals and most small birds, pairing does last for this period, and more than one brood or litter may be born during a single breeding season. Penguins are among the birds that pair for life, even though breeding only occurs at distinct times of the year and the young grow up quite quickly. In man the period before the young are independent is so long that, on purely biological grounds, pairing is desirable for fifteen years or more.

Because the long-term pair bond has to endure until the offspring are capable of surviving independently, it must not collapse. If it does, then the lives of the young, and hence the genetic investment of the parents, will be threatened. Yet the stimulus to keep the bond intact does not come entirely from that investment, although the juvenile expressions and behaviour of the young do, to some extent, inhibit aggression in the parents. The real power holding the parents together for a long time in species such as man is sexual attraction. For this reason man and some of his anthropoid relatives do not have a specific breeding season. Mating takes place all the year round despite the fact that the female is only fertile for a brief period each month.

To maintain a lasting bond both male and female must be able to receive a reward for their fidelity. For the male the prize is the continual sexual readiness of his mate; for the female it is her sexual response to the genital display continually broadcast by the male. In the chimpanzee the large scrotum housing the testes provides the main signal that arouses the female, while in man it is the penis. The orgasm experienced by the female, which is apparently unique to the higher primates, is probably also part of her prize. Both these goals keep the pair together and in harmony for many years.

Bonding is not used exclusively to keep a male and female together. It also unites animals in herds, schools and hunting cooperatives. In such social relationships, great use is made of greetings which have the exclusive function of appeasing aggression and reducing tension. Sometimes greetings between two animals are so complex that they are akin to diplomatic protocol.

Greetings are built upon one of three main sorts of behaviour—threat, parental care and sex. Greetings that have their base in threatening behaviour always have a safety fuse built in. Storks, for example, which clap their bills in aggressive displays, briefly point their bills backward during greeting. Without this quick backward flick, which signifies that the threat is not real, the greeting is turned into an attack.

The postures of parental care are part of many greetings. Touching and stroking play an important part in the pacifying greetings of many monkeys and apes. Old and dominant males will allow themselves to be caressed by subordinate youngsters and will themselves caress both males and females. All this behaviour enhances social harmony within the group. Feeding is another vital parental activity that appears in greetings. In the normal courtship of black-headed gulls, for example, one partner begs from the other and may utter cries more reminiscent of an infant than an adult. Primates—including man—often kiss when they greet. Kissing originates in the passing of chewed food direct from a mother's mouth to that of her infant. This element of juvenile behaviour has received rather special evolutionary treatment in man. For him it has come to play a new role in sexual behaviour.

When two male baboons greet each other, one presents his rear end to the other in much the same way as a female biologically ready for mating exhibits herself to the male. The mimicry of sexual attitudes in this greeting is so complete that the thickened areas around the male baboon's anus are slightly pink in colour like those of the female. A dominant male may briefly mount a subordinate who is presenting his rear, so reinforcing his position in the social scale. Homosexual mounting is quite common in troop-dwelling primates, and serves to keep hierarchies in order.

Animals that have to hunt cooperatively, such as the hunting dog, have evolved a complex group structure which depends for its existence on the continual sublimation of aggression. Man also comes into this category, and although it is always tricky to separate aspects of human behaviour that have a genetic basis from those imposed by the dictates of society, many of man's social greetings do seem to have well-founded biological origins. A universal feature of human greeting is smiling, an innate behaviour shown by infants from their first few weeks of life. The exact origin of smiling is obscure but it may have evolved from one of the gestures used in grooming behaviour. Since grooming and being groomed require complete mutual trust, it is likely that during evolution the gesture of the smile has been transformed into a signal of non-aggression. A smile's effect is as instantaneous as it is disarming—its message is fundamental and powerful.

From the handshake to the close embrace, most human cultures show some sort of stylized physical contact when two people meet. Often this is coupled with an expression of submission, be it a bow or just a nod of the head. Physical contact is a powerful feature of parental behaviour which is studiously avoided in other dealings with strangers. If one accidentally touches a stranger a profuse apology is made in an attempt to appease the aggression that the 'attack' has evoked. No such apology is necessary when the contact is formal and surrounded by ritual.

THE BOND

Natural hostility can be so strong that a courtship can end in the death of the male. But the suspicions must be overcome—at least long enough to sow the seeds of new life.

Greeting ceremonies

Animals are naturally wary of one another. Potentially, two members of the same species are competitors for food, territory or nesting sites. But when a pair of individuals has to spend much time in close contact—as they must in mating or rearing their young—it is essential for them to have a means of dispelling their natural fear. This they do through greeting ceremonies. The natural apprehension induced by the presence of another creature instinctively generates feelings of aggression and/or fear, but greeting behaviour serves to appease the desire to fight or run. Thus the pair may get on with the job of reproduction.

Greeting movements are really a patchwork of small behavioural scraps gleaned from many functional contexts and sewn together to create a fabric of fascinating complexity. Parental care, sexual behaviour and aggression itself form the foundations of greetings. In their true expression these behaviours fulfil a real function, but in greetings they have been translated into rites—

brief pantomimes that must be acted out to make bonding possible. The feeding ritual of incubating cormorants, for example, is parental care in disguise.

Natural selection has ensured that the original purposes of out-of-context greeting behaviour have been effectively neutralized. In a piece of behaviour that has its origins in parental care, this modification is of little importance, but if greeting is derived from threat behaviour it is vital. Thus horses greet one another by opening their mouths and drawing their lips up and back to expose their teeth. This is almost the way in which they threaten each other. The difference is that in greeting displays the ears are held erect instead of back against the head, so cancelling the underlying aggression.

The male greylag goose, *Anser anser*, in greeting a potential mate, threatens inanimate objects, rushing at them with his head low. He returns to his partner and threatens just beyond her. At this point the female joins in the mock battle and both goose and gander lower their

heads and cackle. This cackling has developed from the contact-seeking call of the young, and its presence in the greeting is vital to defuse the aggressiveness inherent in head lowering.

Nowhere are greeting ceremonies more complicated than in man. Many human greetings contain elements acquired from the cultural environment, but many others have their foundation in nature. When analyzed, several of man's greetings are found to have origins similar to those of other animals.

A smile is a universal human greeting. Babies do not need to learn how to smile—blind and deaf babies do so as readily as normal ones. A gesture of submission—be it little more than a nod of the head—is a greeting designed to appease aggression. Hand-to-hand contact with the bare hand perhaps indicates the absence of a weapon and the desire for close contact. The social convention of taking a gift on visiting a friend's house may be similar to the cormorant's ritual presentation of seaweed to its mate.

A white stork, *Ciconia ciconia,* may spend hours, even days, sitting on its eggs. When its mate comes to relieve it, they perform a greeting display. The sitting bird rises and flicks its head round over the shoulder. It 'claps' its bill rapidly while bringing its head back to the front and then raising and lowering it. The moment when the bill points back defuses any fear of threat. The arriving bird joins in and the ritual may be repeated several times.

When a flightless cormorant returns to the nest to relieve its mate sitting on the eggs, it brings a beakful of seaweed. The bird offers this gift to the sitting partner who jerks it from its mate's beak. A bird arriving without a gift is driven away by its mate. Bringing a gift diverts the latent aggression of the sitting bird. The flightless cormorant, *Nannopterum harrisi,* lives in the Galapagos islands.

GREETING
Physical greetings
are an important
feature of the life of
the chimpanzee, *Pan
troglodytes*. When
the troop is moving
the chimps often
shake hands, and
when at rest they
touch each other's
faces. Both actions
are derived from
behaviour between
mother and young
and have elements
of begging for food.

A full embrace
between adult
chimps is the closest
type of greeting.
This action, often
seen between an
adult male and
female, stems
directly from the
infant clasping
behaviour seen in
all apes, and man.
Often, the embrace
is solicited by a
subordinate seeking
the approval of a
dominant individual.
Even old males will
attempt to embrace
younger males.

Staying together

To retain the friendly relations initially established by greetings, animals must adopt submissive, non-aggressive postures whenever they can. Like the rituals of greeting, the elements of this submission are often borrowed from other sorts of behaviour and transcribed into a new language.

Whether they have scales, fur or feathers, most vertebrates need to groom and preen themselves to rid the skin covering of fleas, lice, ticks and similar parasites. But some parts of the body, such as the back of the head and the neck, are hard to reach, and those sites have become the focus for the change in function of grooming behaviour. In getting another animal to attend to the hygiene of these awkward parts, the individual receiving the treatment must lie down or hold its head low.

In the evolution of behaviour it is hardly surprising that grooming and preening postures have become acts of submission. They are used as such to pacify aggression. The comfort and reassurance given by localized stimulation of the skin is universal. If, for example, the play between young rats or rabbits becomes too rough and one partner is accidentally bitten, the loud squeal of the injured party will immediately invite the 'aggressor' to groom his playmate. Such behaviour has its parallel in humans—when a toddler falls down and knocks his head his mother will 'rub it better'.

Because the readiness to groom signifies peaceful intentions it also forms a vital part of some greeting ceremonies. Lemurs groom by combing their fur with their lower incisor teeth, which project forward in front of the face, and condition their fur by licking it. During greeting displays, friendly intentions are made obvious by mock combing and rapid licking movements. Similar behaviour is also seen in macaque monkeys.

Among the social birds and mammals, preening and grooming do much to ensure good community relations. Usually the socially superior individuals solicit grooming from their inferiors, reinforcing the fact that in a structured society, harmony is not only possible but desirable. To curry favour, social inferiors clamour for the privilege of picking over the fur or feathers of their superiors. The recipient of this attention does not have to put on a brave face to endure the grooming; on the contrary, preening and grooming induce a state of great pleasure and relaxation. Such behaviour does much to enhance close physical contact between the members of a population and thus to improve social integration.

● Bird 1

● Bird 2

Singing together is one way in which birds maintain a pair bond. In some birds, although both male and female are capable of singing the whole song of their species, they divide the phrases of it and sing alternately. So perfect is their timing and integration that the human listener finds it almost impossible to discern the exact moment of changeover. The mutual dependence that this habit brings keeps the partners together until their chicks are safely raised and have left the nest. This type of bond maintenance is found only in the densest forest habitats. African boubou shrikes, *Laniarius aethiopicus*, left, are among the birds that perform this type of duet. Others are the tyrant-flycatchers of South America, Central American wrens and grass warblers.

Grooming is an important and time-consuming activity. The groomer rakes through the fur with his fingers to remove large dirt particles, pieces of vegetation and fleas. Ticks and mites, however, are removed with the teeth. The animal carefully closes its mouth round the embedded sucking neck of the parasite and eases it out. The closeness of one monkey's face to the skin of another, as in the vervet monkeys, *Cercopithecus pygerythrus*, below, also helps later recognition by smell.

Birds preen their feathers by different methods. The most efficient action is to use the end of the beak to peck the feathers from base to tip. This removes parasites, loose feathers and stale preen oil. A second method is to grip the feather at its base and 'strip' it in a single movement. Also, with closed beaks, birds rake through their feathers to repair breaks.

Penguins mate for life, and in species such as the Humboldt penguin, *Spheniscus humboldti*, both parents assist in incubating and rearing the young. The male may have to sit on the eggs for a fortnight or so, awaiting the return of the female from a post egg-laying feeding trip. Social preening plays an essential part in the changeover ceremony at the nest. Head, neck and chest regions are particular targets for preening, and often the penguin partners indulge in mutual grooming.

Countering female hostility

The act of mating is often the culmination of the long process of courtship and pair formation. For many marine and freshwater creatures, however, it is a simple, perfunctory act: they simply release eggs and sperm into the water so that fertilization takes place externally. The water protects the eggs and sperm and brings them together for fertilization.

Most marine polychaete worms use this method. In both sexes the sexual products are formed in segments at the hind end of the body. As the eggs or sperm develop, fleshy flaps, parapodia, on the hind segments enlarge and flatten. The worm darkens in colour and its eyes grow larger. When these changes are complete, the worms, which usually live hidden among rocks and stones, come out and swim freely, using the enlarged parapodia as paddles. The wall of the hind segments of each worm splits and the contents are released into the sea, where some of the eggs will be fertilized by sperm.

Some marine invertebrates do have methods of internal fertilization. Octopus and squid, for example, have a special arm with which they place bundles of sperm into the mantle cavity of the female. The octopod, *Argonauta*, has a particularly interesting mating arm: the end carrying the sperm detaches and remains inside the female.

Internal fertilization of some type is essential for all animals that mate on land, as the eggs and sperm cannot survive long outside the body. Some insects, such as scorpions, place spermatophores, sperm enclosed in capsules, on the ground, where they are then picked up by the female. Most land-dwelling invertebrates, however, must deposit their sperm right into the female.

Many, such as mosquitoes and dragonflies, have copulatory apparatus at the end of the abdomen. The male flea has a complicated organ which must be exactly the right shape to fit into the female's equally complex aperture. Each species is slightly different, thus providing a 'lock and key' barrier to hybrid matings between species.

For birds and mammals, the final stages of courtship serve to quieten and placate the female before copulation. In solitary carnivores, particularly, the male must bide his time until the female's aggression is suppressed by her sex drive. Strangely, many female insects and spiders seem unable to control their predatory natures even for the few moments needed for mating. In such cases the male counteracts the female's hostility in various ways. The male balloon fly offers a gift of prey to the female who becomes so engrossed with it that she does not notice that she is being mated. The male tetragnath spider holds the female's fangs so she cannot attack; another male spider ties the female's legs down with silk.

The mating procedure of the praying mantis is most bizarre. The male offers himself to the female, who promptly bites his head off. Losing his head actually stimulates the male to copulate. Furthermore, his body provides some of the necessary protein to nourish the developing eggs inside the female.

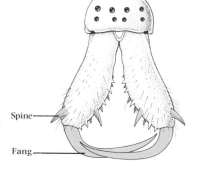

The fangs, chelicerae, of the male hunting spider bear hard spines. They are positioned so that the male can keep the female's lethal fangs apart when she attacks. The action of the male's huge fangs pressing downward on the female's jaws locks the pair tightly together.

Spine

Fang

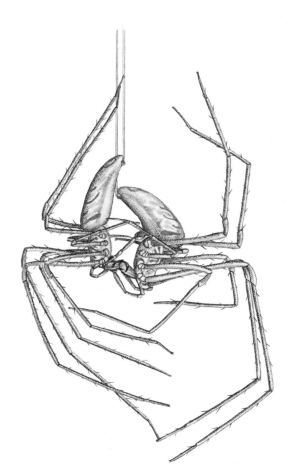

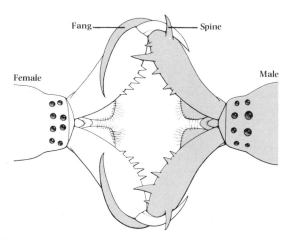

Fang — Spine

Female

Male

Male hunting spiders, *Tetragnatha* spp., cannot calm the aggression of the female of the species. In order to mate with safety they make use of their elongated fangs, chelicerae, equipped with hardened spines.

Before the male sets out, he builds a tiny platform of silk, deposits sperm on it and dips his palps into the sperm. Once he finds a female the male stands his ground while she attacks. He deftly moves his chelicerae

so that the spines wedge her fangs open and prevent her biting. In this position he inserts his palps into the sperm storage organ in her abdomen and fertilization is complete.

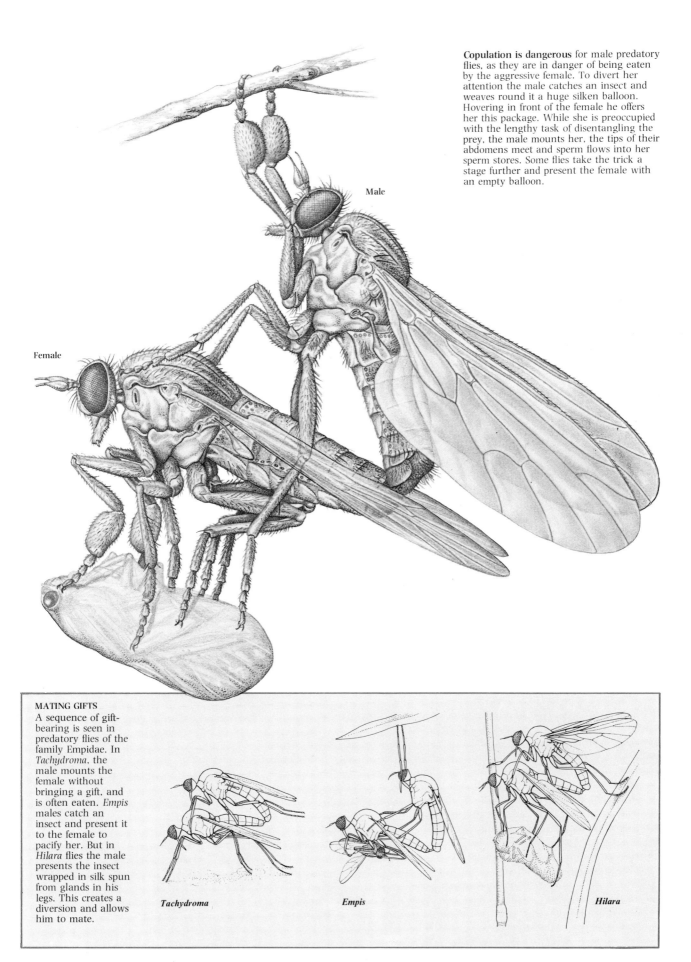

Copulation is dangerous for male predatory flies, as they are in danger of being eaten by the aggressive female. To divert her attention the male catches an insect and weaves round it a huge silken balloon. Hovering in front of the female he offers her this package. While she is preoccupied with the lengthy task of disentangling the prey, the male mounts her, the tips of their abdomens meet and sperm flows into her sperm stores. Some flies take the trick a stage further and present the female with an empty balloon.

Male

Female

MATING GIFTS
A sequence of gift-bearing is seen in predatory flies of the family Empidae. In *Tachydroma*, the male mounts the female without bringing a gift, and is often eaten. *Empis* males catch an insect and present it to the female to pacify her. But in *Hilara* flies the male presents the insect wrapped in silk spun from glands in his legs. This creates a diversion and allows him to mate.

Tachydroma

Empis

Hilara

The mechanics of mating

For all reptiles, birds and mammals, the fertilization of the female's eggs by the male's sperm must take place deep within the female's body to ensure the development of the embryo. Among land-living vertebrates various mechanisms have evolved to guarantee that sperms are deposited as far inside the female as possible.

Reptiles, birds and the lowly mammals (the marsupials and egg-laying monotremes) have a single exit to their bodies. This is the cloaca, a chamber into which the tubes from the urinary, reproductive and digestive systems lead. In reptiles and birds the cloaca plays an important role in mating. Most male reptiles do not have true penises. Instead they have two curved hemipenes which can be erected forward and outward. Only one hemipene, the one closest to the female's cloaca, actually penetrates her. To erect the hemipenes from their relaxed positions two mechanisms are used, one muscular, the other hydraulic.

The reproductive anatomy of a male bird is rather different. He is equipped with a small, erectile papilla at the base of his cloaca through which sperms pass during mating. This papilla penetrates the female's cloaca just far enough to reach the end of the oviduct, the tube down which eggs travel from the ovary. During mating strong muscular action opens out the cloaca of both male and female. Mating is a very brief 'cloacal kiss' in which the male perches precariously on the back of the female.

Deep fertilization is made possible for mammals by means of the male's rather large penis. Many mammals carry their penis retracted into the body, but it can be erected at the moment of mating for insertion into the female's vagina. Erection is effected by hydraulic action alone, as in man, or by combining hydraulic and muscular mechanisms. In most mammals—but not man— the penis receives extra support from a bone, the baculum or os penis, which

runs along its shaft. In shrews and some other insectivores the penis is so long that the baculum is hinged in the middle and the penis folds up when not in use. In some of these creatures the tip of the penis or glans penis is decorated with rows of hooks and spines. These probably prevent the penis from slipping out, but in species for which the friction of copulation is essential to induce the release of eggs by the female the spines may provide added stimulation.

To mate, most terrestrial male mammals stand up and enter the female from behind. The penis is inserted for only a second or two in rodents, but in the rhinoceros intromission lasts half an hour or more. Carnivores often collapse during intercourse and remain joined for many minutes after transmission of sperm is complete because the tip of the penis becomes enormously swollen, making withdrawal impossible. This prolonged union gives the sperm the best chance of reaching the egg.

THE FEMALE'S VIEW
Genital display and size is linked to the strength of the pair bond necessary for rearing the young. To the female's eyes, the male gorilla has a tiny penis and tiny testes; the chimpanzee has a larger penis and huge testes; man has relatively large genitals for his size. For man and the chimp, the pair bond is reinforced by sexual activity.

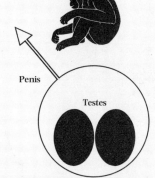
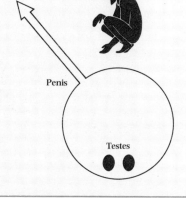

Gorilla — Penis — Testes

Chimpanzee — Penis — Testes

Man — Penis — Testes

A bone, the os penis or baculum, lies inside the penis of many mammals and gives it the necessary rigidity for copulation. At its base it is held by ligaments to the pubis. The tip may form fingerlike projections, a hook, or a rough palmlike structure. The os penis of the sea lion is about 2 ft (60 cm) long and weighs 4.4 lb (2 kg) while that of a bull is almost 3.3 ft (1 m) long. The tip of the bone can be used to distinguish between closely related species, as within a species its shape is identical.

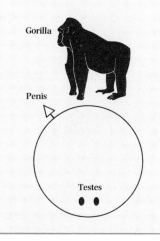
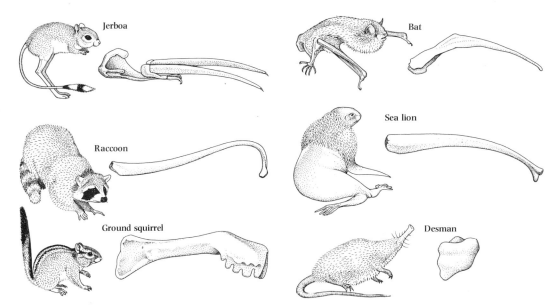

Jerboa

Bat

Raccoon

Sea lion

Ground squirrel

Desman

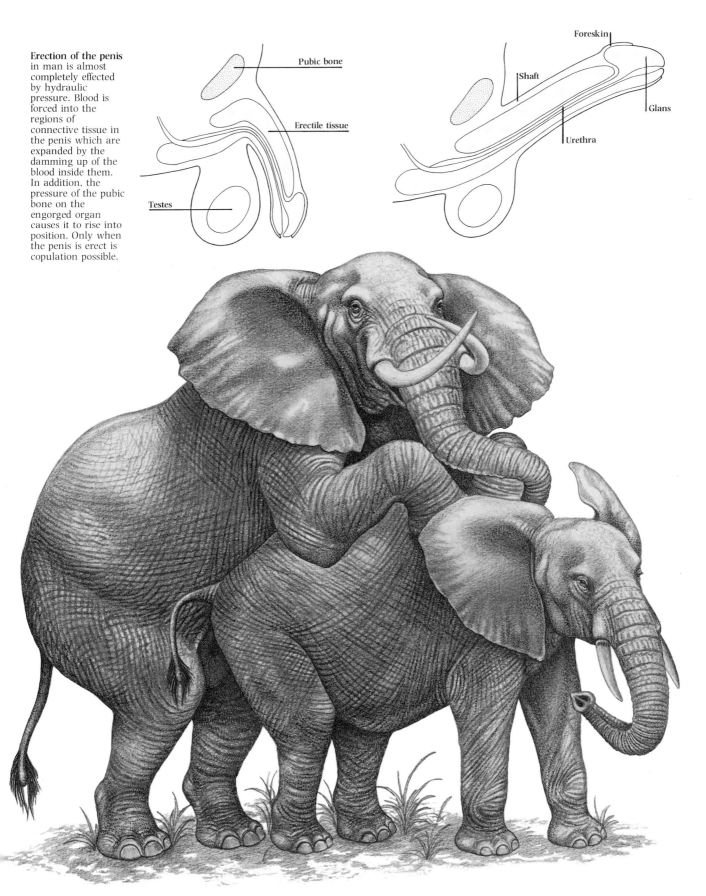

Erection of the penis in man is almost completely effected by hydraulic pressure. Blood is forced into the regions of connective tissue in the penis which are expanded by the damming up of the blood inside them. In addition, the pressure of the pubic bone on the engorged organ causes it to rise into position. Only when the penis is erect is copulation possible.

Pubic bone

Erectile tissue

Testes

Foreskin

Shaft

Urethra

Glans

Elephants mate from the rear, as do most mammals. The penis is extremely long because of the huge bulk of the creature—it measures almost 5 ft (1.5m) when fully extended and weighs 60 lb (27 kg). Even so, little more than the tip of it actually penetrates the female's vagina. The mating of elephants is an ungainly affair, with the male making repeated efforts to support his weight on the female's back. Only whales, man and, on rare occasions, chimpanzees, orang-utans and gorillas mate face to face. All animals are vulnerable to attack while in the mating position but, because of its size, the elephant has little to fear.

Hierarchies develop among the majority of vertebrate animals and certain invertebrates. In some species the prize for the most dominant members is higher social rank and hence the right to the best food or best mate. But in many other species it is the possession of a tract of land. The behaviour associated with an animal's attempts to secure and defend such an area, or territory, is often curious and bizarre. Although all animals place great value on the establishment of a territory, its function is not the same in all species. There is little doubt, however, that territorial behaviour exists to ensure that the population never grows too large: only territory holders breed. Overpopulation leads to over-exploitation of food and other resources, and consequently threatens the survival of all.

Charles Darwin pointed out that animals have the capacity to multiply beyond the number of adults that the environment can accommodate. In his lifetime, predation and climatic conditions were regarded as the regulating forces preventing overpopulation. Today it is believed that while these factors influence population numbers and have a major effect on the lives of certain invertebrates, much of vertebrate behaviour is designed to establish and maintain a rigorous class structure in which only the privileged have the right to breed. Often, but not invariably, the underprivileged do not get enough food and soon become diseased and die. In animal societies the members compete for privilege in the form of high social status or territory.

The mechanism of territorial behaviour is basically the same in all species, regardless of the specific function of the territory. It is usually the males who take territories, though in some carnivorous species the females also do. In general, the establishment of a territory takes place in three stages. First, two adult males confront and threaten one another with their weapons or finery, for in the animal kingdom the male is usually more highly adorned than the female. While it is true that male finery serves to attract the female at the outset of courtship, it also plays a substantial role in the rivalry between adult males. Such competition, however, rarely results in injury to either combatant. The territorial boundary disputes that break out are carefully planned displays capable of intimidating the rival without violence. In small garden songbirds it is the splendour of the song that intimidates; in the mouth-breeder fish it is the intensity of coloration and the complexity of the gill-cover raising display; in deer and antelopes it is the strength of one of the contestants. What is particularly interesting among the horned or antlered species is that the competitors do not charge at each other as soon as they meet; instead, they engage their horns or antlers and commence to push only when they are certain these will not slip apart. The ridges on the horns of most antelopes are designed to prevent slippage, which can result in accidental injury to the opponent's body.

The second stage is when the loser gives in gracefully. He may do this by turning around and moving off, or he may behave in a way that will appease the victor's aggressiveness. Either way, the defeated animal is free to leave and compete with another male elsewhere. In a few species, such as the kob antelope of Uganda and the shelduck, unsuccessful males form herds or flocks. As a rule, they are relegated to the poorest habitats. Studies of the red grouse, which inhabit the highlands of Scotland, have shown that the unsuccessful males trying to scratch a living from marginal habitats are the staple diet of foxes throughout the winter. In terms of population numbers they are relatively expendable, for only territory holders breed.

The third stage begins when the successful male tries to protect what he has won. Some species, such as birds, sing to defend their own area, but other species resort to leaving signs around the frontiers. Many mammals mark their territories with secretions that are produced by specialized scent glands, thus informing would-be pirates that the area is occupied. Urine and faeces are also used as scent marks. Male dogs lift their legs to urinate as high up an object as possible. Bitches, on the other hand, do not defend a territory and so do not show leg-cocking behaviour. Carnivores constantly patrol their territories, renewing their twenty or thirty scent marks every few days. Marks are made as obvious as possible so that the neighbours and wanderers that must be able to find them can do so quickly and reliably. Faeces, together with the gelatinous contents of glands close to the anus, are almost always placed within a few yards of vertical objects such as trees and fenceposts.

Studies of rabbits reveal that scent marks are not left exclusively for the benefit of other rabbits outside the territory. A rabbit within its own territory and surrounded by its own odour acts in a confident manner and wins all its territorial encounters. Removed from the territory, it acts in a downcast manner and is easily beaten by much weaker opponents. In addition, territorial boundary marks do not invariably say 'keep out'. They serve to educate and inform, and contain information about the age, social rank, sex and perhaps even size of the tenant. After considering these facts, the visitor can then decide if he has anything to gain by penetrating farther and risking a fight.

Though it is usual to find a single male taking a territory, this is not always the case. In many gregarious species, groups defend a territory. Howler monkeys, in groups of up to twenty, patrol a territory and shout their threats to neighbouring groups. But they establish few new territories, since ownership of the defended area is passed on to others in the group when the oldest individuals die or are killed. Another variation is seen among burrowing mammals. Their defended territory, or core area, is only a small patch around the burrow entrance. But in the home range, which is a much larger area that spreads outward from the core area, the tenant has the right to move about freely. Other tenants of other core areas may also use the home range, or part of it. Avoidance, rather than conflict, is the rule here.

In a number of species, territorial behaviour is strongly influenced by the time of year. Carnivorous species hold territories throughout the year, and scent demarcation is a daily chore. Although blackbirds occupy territories all year round, the defence of the winter territory rests solely with the female. Many deer and antelopes take territories only during the breeding season; in fact, this is the commonest time for most animals since all territorial behaviour is more closely linked to the social than to the physiological side of reproduction. After all, unsuccessful territorial males are as sexually capable as successful males.

The primary function of a territory is to provide adequate food for its tenant and his family. If prey is scarce, the territory must be larger than if it is abundant. Some carnivorous territorial birds, such as owls, have evolved a sophisticated means of regulating their reproductive output to match the food resources. Owls always lay four eggs, but with two-day intervals between each one. Since they hatch sequentially, the first chick hatched is older, and therefore larger, than its siblings. If the territory is rich in food, all four chicks survive; if the supply of mice fails, the larger chicks will trample on their weaker nest mates in competing for the attention of a food-bearing parent.

Some territories are for mating purposes only. There are male birds, such as bowerbirds or manakins, which display among themselves for small territories in which they build beautiful structures to attract females. If they are able to lure a female into the bower, courtship and mating take place. But then the female flies off to build a nest elsewhere. Lekking, a specialized form of territorial behaviour in which males compete for a tiny patch of a traditional breeding site, is similar. The territory of the kob antelope, for example, contains neither food nor water, and the tenant male must leave the site after a few days to regain his strength in lush feeding grounds nearby. Lekking game birds usually compete for a couple of hours a day before flying off to feed in a communal flock.

In sea birds, particularly gannets, boobies and their relatives, territories are for nesting only. Mating takes place away from the nest site and competition between females separates one nest from another. It is difficult to understand what prevents some birds from opting out of the territorial system and nesting or mating in places where competition is lacking. What is at stake, however, is more important than any short-term gain, for the survival of the whole population would be put in jeopardy if individuals broke the rules of conventional behaviour. Territorial behaviour, whether it is for a tract of land supplying food resources, or specific and identifiable mating places, or nesting sites, provides a powerful social filter in which the prize is the right to breed. Since those individuals that are successful have obeyed the rules, their offspring carry the same genes of obedience, and thus the trend is continued.

TERRITORY

First, a space has to be won that will provide enough food for the prospective family. Then it must be staked out and protected. For successful tenants the reward is the right to breed.

Ritual battles

An animal with a territory has to maintain and defend it by putting would-be usurpers firmly in their place. This aggression is not in any way related to the type of behaviour exhibited by a predator attacking its prey. The motivation in aggression between members of the same species (intraspecific) is social; the individual desires to possess what is necessary so that he may attract a mate.

In aggression between animals of different species (interspecific) the motivation comes from the hypothalamic portion of the brain that is associated with the hunger drive. Thus male and female predators have to express this type of aggression if they are to survive, but normally only the male indulges in the ritualized aggression involved in defending territory. In fact, many animal experiments indicate that the strength of this aggression depends upon the amount of male sex hormone, testosterone, in the blood. The distinction between the two types of aggression is further proved by the fact that nonpredatory, grass-eating species, such as the oryx, show just as much territorial aggression as do predatory species such as the rattlesnake.

All intraspecific fights are stylized. In some species, such as the Siamese fighting fish, the fight is similar to a ballet in which each step is carefully worked out beforehand. It is vital for the well-being of the species that the contestants suffer no physical damage as a result of the encounter. Consequently, natural selection has ensured that weapons and threatening behaviour are specialized in such a way that they focus the attention of the contestants upon their ritual performance. The horns of most antelopes, for instance, are ridged so that when competing males engage theirs, the risk of the horns slipping off one another and causing injury is minimized. Deer with branched antlers carefully lock them together before pushing against one another.

Generally, the better developed an animal's armoury, the more stylized and ritualized is its aggression. Large effective predators, such as the lion and leopard, use threatening behaviour in which they bare their teeth. They seldom engage in physical contact and the loser displays appeasement gestures when he knows he is beaten.

Since all complex behaviour is built from separate components it is not surprising that many facets of intraspecific fighting are seen in courtship. Antelopes play with their horns, raising or lowering them according to the stage in courtship. The message conveyed by a particular type of behaviour depends upon its context in the entire sequence.

Rattlesnake

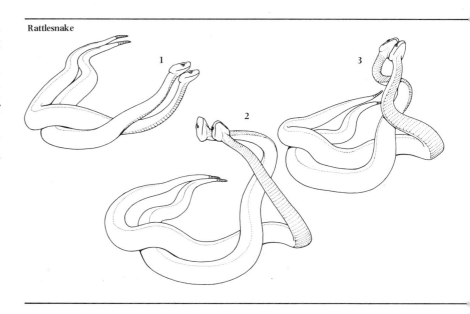

Siamese fighting fish

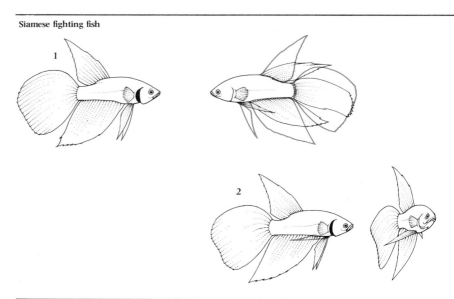

Oryx

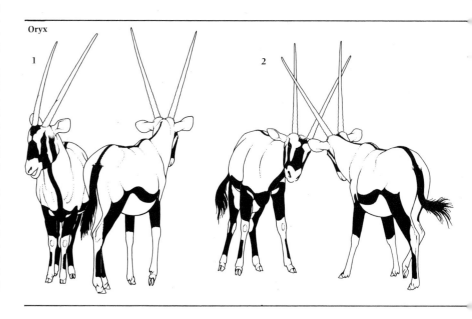

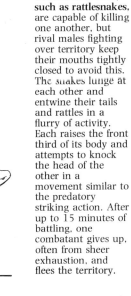

Venomous snakes, such as rattlesnakes, are capable of killing one another, but rival males fighting over territory keep their mouths tightly closed to avoid this. The snakes lunge at each other and entwine their tails and rattles in a flurry of activity. Each raises the front third of its body and attempts to knock the head of the other in a movement similar to the predatory striking action. After up to 15 minutes of battling, one combatant gives up, often from sheer exhaustion, and flees the territory.

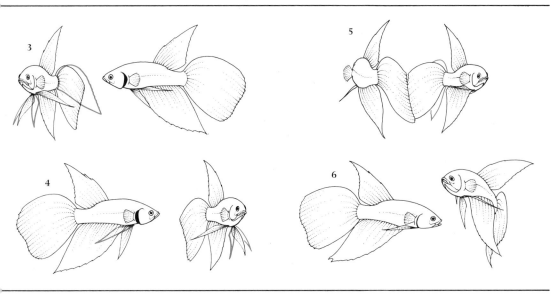

The essence of the battle of Siamese fighting fish, *Betta splendens*, is threat. The fish extends his fins fully and opens his gill coverings until they stand out at right angles from his head. The display is usually kept up for a couple of minutes, briefly released and then resumed. The fish makes repeated attacks on his rival but usually stops short of biting him. Instead he directs his aggression into an even more fantastic fin display. If his opponent fails to submit, a fight may cause injury.

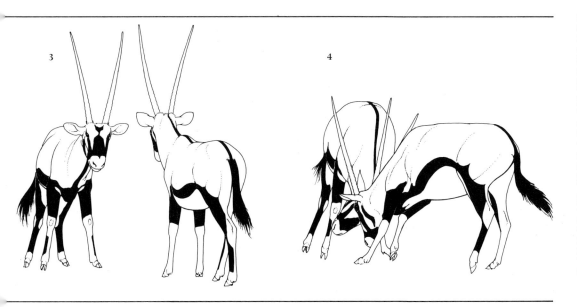

Gregarious antelopes, oryxes, *Oryx gazella*, do not defend territories, but males fight for possession of a group of females. The scimitarlike horns can be weapons but the oryx does not use them in this way when fighting a rival. The males cross their horns and push a little. They break apart and begin head-to-head wrestling. This struggle may continue for many minutes and be resumed time and again until the superior power of one forces the other to give up.

Marking the boundaries

Demand for territory is so great that an animal must advertise to other males, and females, that he is in residence. Otherwise, he will soon be deposed. Visual signs are useful only at close range, so animals with large territories use sound or scent signals. Hyenas deposit faeces at prominent sites on their group boundaries; these are easily spotted by other hyenas which then sniff them for specific information. Although acoustic signals can be produced at night or during the day, they require the producer to be present. Olfactory signs, once deposited, broadcast their message by day and by night and have the advantage of continuing to work long after their producer has gone away from the area.

Most territorial birds advertise their presence only during the early mornings in the spring. They fly to a high point in their territories from which they can see and be seen. The rest of the day is spent in nest building, courtship or fetching food for the hungry young fledglings.

Mammals, however, do not divide up their day in this way and their territorial demarcation is a continuous process. A number of species have developed specialized urination and defecation behaviour to mark their territories. Dogs lift up their legs to urinate as high up a tree or other object as they can. This gives the maximum possible surface area for evaporation. Badgers clear a small area and deposit their faeces in the middle, thus making them more obvious. Wombats leave their faeces on tree stumps, rocks or other raised surfaces to make their presence known to others.

The majority of mammals mark their territories with secretions from specialized scent glands. In the rabbit, the scent gland lies in the angle of the lower jaw, just beneath the chin; a delicate chin-rubbing action smears the scent on sticks or rocks. Species with glands on the hips or flanks, such as the water vole, use their hind feet to scratch the glandular surface and then scent the ground by stamping. Weasels, stoats, otters and a few rodent species have scent glands in or near the anus. They leave their mark by dragging their back ends along the ground. One of the most unusual scent-marking behaviours is shown by the marsh mongoose, *Atilax paludinosus*, of East Africa. It applies the secretion of its anal glands to overhanging branches by adopting a handstand posture and then transferring the odorous secretion.

Territorial boundaries are not untouchable; they serve as information posts about which animals own what and various information about them. Most territory holders will only attack an intruder that threatens the exclusive core area of its home.

With delicate head movements, the black-buck, *Antilope cervicapra*, marks a branch with a secretion from a gland beneath the eye. The secretion is sticky, and a large drop is usually applied. The smell is attractive to others of the same species; they sniff it for long periods before overmarking the same spot or marking an adjacent twig. Many other types of antelope possess this scent gland.

The scent gland of antelopes lies in a bony depression on the face in front of and beneath the eye. It is modified from skin glands and much of its structure resembles that of the oil-producing sebaceous skin glands. It also resembles in part the sweat glands which produce a watery secretion.

Detail of gland

Lid

Opening

Scent gland

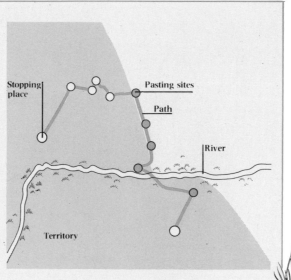

HYENA TERRITORIES
Hyenas are gregarious animals. To mark the boundaries of their group territory they position large masses of faeces close to its edges. Hyenas also use a secretion from their anal glands to mark boundaries—a process known as pasting. The animal sniffs a grass stalk then moves forward dragging the stalk between its legs. As the grass is pulled past the anal opening it is pasted with a creamy white secretion.

Stopping place

Pasting sites

Path

River

Territory

The incredible voice of the howler monkey is due to a specialized larynx. Its major adaptation is the bulbous expansion of the hyoid bone which forms a resonating chamber for the sound produced. To accommodate the structure, the angle of the jaw is spread sideways and deepened to provide protection for this bony balloon. A thick beard obscures the bulge.

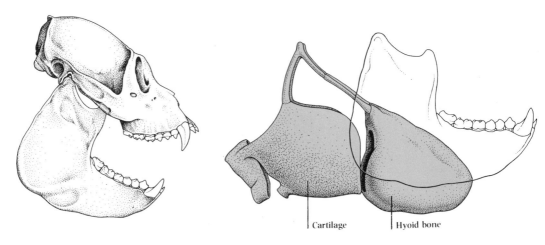

Cartilage   Hyoid bone

High among the tree tops of the tropical rain forests of Central and South America, howler monkeys live in groups of up to 20 animals. Each group defends a territory. To do this the monkeys shout incessantly at the other, rival, groups. If they were to stop shouting it would signify the end of their occupancy. The loudness of the voice is directly proportional to the size of the territory. Northern species that have abundant food and smaller territories have quiet voices, but monkeys with larger territories have powerful voices. The shout of the red howler, *Alouatta seniculus*, right, can be heard from over 3 miles away.

Winning the right to breed

The traditional meeting site, or lek, where adult males gather year after year to display to females, provides none of the requisites for life. Instead, it functions as an alternative to overpopulation. In territorial species, the amount of land is limited—if food stocks are low, territories are large; if they are high, the defended area is small. The number of territories in a particular area regulates the amount of breeding, since only territory holders breed.

All members of a lek system are obliged to obey its strict rules. Although there is nothing to stop an individual from leaving the traditional site and mating in secret, the initiative would be harmful to the population as a whole. As the animals in the group are usually related to some degree, they will have genes in common. Thus, the apparent altruism seen in those males that accept defeat is merely the means by which the genes in the population defend themselves against overpopulation and ecological ruin.

Males of the Uganda kob antelope, *Adenota kob thomasi*, show a similar type of behaviour. Each one competes for a small number of patches of barren earth, about 150 feet (45 metres) in diameter. These are used for display and mating only; the animals have large peripheral territories for feeding. Blood is seldom drawn in these ritual battles when two contestants pit their strength against one another. Finally, the weaker male gives in and leaves the territory. But even the victor can only stay there for a couple of days without food or water. He will strut around, head held high, waiting for a female to enter his territory. If she does, he will do everything to keep her there. If she is more attracted to a neighbouring male, however, he will make no attempt to follow her. Once a female has been mated, she leaves the lek and returns to the female herd to undergo her pregnancy and rear her young. Male herds contain the young and vanquished. They stay together until they grow strong enough to be capable of winning a territory and, with it, the right to breed.

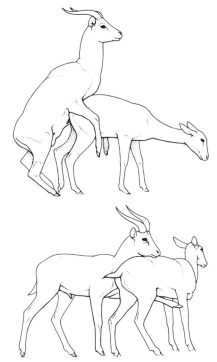

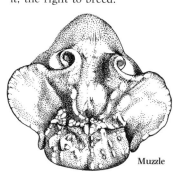

Muzzle

Once a female is attracted into his territory, the male Uganda kob starts a gentle and complex ritual mating. He mounts the female and copulates once. Afterwards, he lays his head on her back and squeezes her belly with an upward movement of his front leg. He holds the female in this pincer posture for a moment then releases her.

Bats are gregarious by nature, but true lekking behaviour is seen only in the hammer-headed bat, *Hypsignathus monstrosus*, of western Africa. At dusk up to 100 males gather at a traditional site in the trees. They set up a deafening chorus of a harsh noise best described as a 'pwok', and for a couple of hours all other forest sounds are drowned. Each bat sets up his own rhythm of noises and can make 80–120 sounds a minute. This bat is about 10 in (25 cm) long and has an enlarged muzzle and larynx.

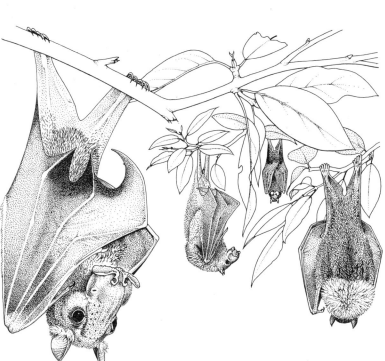

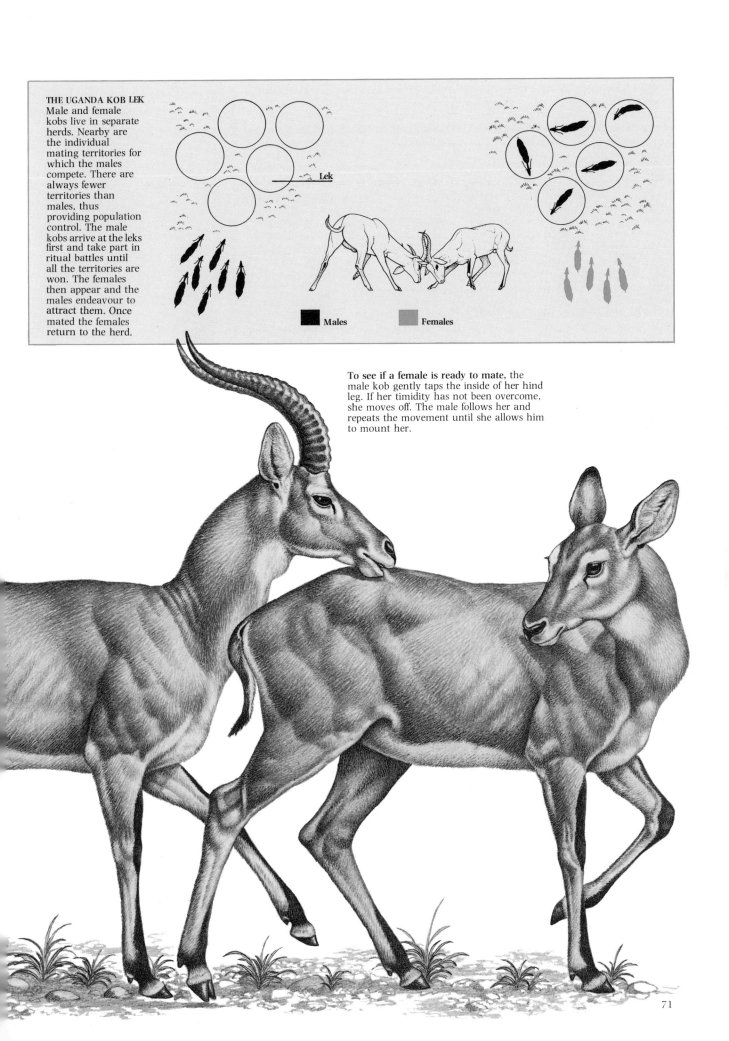

THE UGANDA KOB LEK
Male and female kobs live in separate herds. Nearby are the individual mating territories for which the males compete. There are always fewer territories than males, thus providing population control. The male kobs arrive at the leks first and take part in ritual battles until all the territories are won. The females then appear and the males endeavour to attract them. Once mated the females return to the herd.

Lek

Males

Females

To see if a female is ready to mate, the male kob gently taps the inside of her hind leg. If her timidity has not been overcome, she moves off. The male follows her and repeats the movement until she allows him to mount her.

Status contests

Showy displays performed at specific sites which are used year after year, are the basis of the lek system of attracting mates. Lek displays are used by many game birds such as the prairie chicken, the sage grouse, the sharp-tailed and the black grouse, and by the ruff, a species of Eurasian wader. The males of these species are usually larger, and more brightly and elaborately attired than the females. As with the mammal lek, the system provides a population control, as only a few males, or even only one, may mate at one lek in this strict hierarchy.

All leks function in basically the same way. Male game birds gather at the lekking site in the early autumn. For up to two hours around dawn each day, the cocks puff up the coloured skin bags above their eyes and under their chins, fan out their tail and chest feathers and strut about making mock attacks at one another. But once the sun rises, they put away their finery and fly off to feed. This pantomime is repeated until the breeding season in May. The weakest birds cannot withstand the stress of the lek and gradually drop out of the contests.

By midspring, a dominant cock and one or two subordinates emerge. They hold the territories in the central areas.

When the females visit the lek, which they do only once, up to 85 per cent of the matings are performed by the dominant cock. The number of females mated varies from year to year, presumably in response to the changing food resources. If the dominant bird is removed shortly before the breeding season, social unrest increases dramatically and every mating is interfered with by low-ranking males that are no longer kept in their place. This results in few successful matings.

The male ruffs compete for ownership of small territories, and each victor allows several satellite males to pay him court. Usually young birds, they offer no threat to the resident's supremacy, and are, in fact, an asset because they make his precise location on the lek more obvious to the females. In contrast to female grouse, female ruffs, reeves as they are called, may mate with several males on one lek and may visit several leks within a few days.

THE SAGE GROUSE DISPLAY
The sage grouse, *Centrocercus urophasianus*, of western North America, undergoes a remarkable transformation when it arrives at the display ground. Its tail is raised and the 20 feathers spread out in a fan to reveal their pale undersides. The dark breast feathers are puffed out and the feathers of the nape of the neck lift to form a crest. Finally, a pure white throat sac inflates to such an extent that all the feathers surrounding the face stand outward and upward obscuring the face. In this posture the male grouse shuffle about quivering their tail fans, trailing their wing tips and making curious grunting and purring noises.

Female birds, reeves, are attracted to the display area by the males' fluttering actions. A ruff flies up to about 20 ft (6 m) and then tumbles down, exposing the white undersides of his wings as he falls.

Female

Lek displays take place in open flat areas such as forest clearings. In the autumn, birds display over the whole area, but, as the breeding season approaches, activity becomes localized in a small inner zone where the birds mate. Females arriving at the lek move slowly inward to the copulation zone. Once mated they fly off.

Copulation zone

Ruffs, *Philomachus pugnax*, perform communal displays. Only the territory holder, distinguished by a dark or black ruffle of feathers at the neck, mates. He is surrounded by younger satellite males with pale or white feather ruffles. As a male ruff grows older his head and neck feathers darken. When they are nearly black he may become a resident territory holder.

Of all animals, man is one of the greatest home builders. The human concept of a home is a complex building, or a simpler dwelling such as a tent or a hut, which provides shelter from the harsh conditions of the outside world. The home of modern man has progressed far beyond the concept of a shelter and is more an effort to control the living environment. But primitive peoples such as nomadic hunter-gatherers pause only to construct a simple, temporary roof over their heads in which to sleep and shelter. More sedentary hunters or farmers make semi-permanent homes from whatever material is available—usually vegetation, but also mud or even snow.

Although many animals appear to do very well without a home, many others are accomplished builders, and some make sophisticated buildings in attempts to control their surroundings. To understand animal home building it is important to differentiate between a home and a territory. A territory is an area of the environment which an animal will defend against all others of the same species and which contains most of the resources that the animal needs to live on. A home is a specific place within the territory, which the animal reserves for rest, shelter, protection and possibly for the breeding and rearing of young. This home may be specially constructed, like a bird's nest, or be just a natural hole or a particularly sheltered spot. It may be a simple, temporary affair or an elaborate structure which provides for all the animal's needs.

Many invertebrates, the animals without backbones, never have a home at all. Instead, they spend their lives as permanent nomads, wandering or drifting through their surroundings. In many cases when these animals have mated the eggs are left to develop on their own.

The offspring of many marine invertebrates drift near the surface of the open sea as immature larval forms or plankton. A mature prawn or crab may eventually occupy an area that could be defined as a home, but for shelter merely burrows into the sand or hides in a convenient place when the tide goes out.

In some animals the home habitat starts when one particular hiding place is regularly used. Many of the tropical fish that live in coral reefs rush into their own special crevices whenever a predator approaches. But even such an 'accidental' home has an important advantage—it is a safe place that can be found quickly when there is no time for searching.

A home that is safe for an adult is even more valuable to the small, vulnerable and defenceless young, who may stay home-based until they can fend for themselves and thus stand a better chance of surviving in the harsh outside world. If the young are at home, the parents know where they are and can bring food if necessary. So whenever an animal has a home there is a tendency to breed and raise the young in or near it for safety. Often the home may be selected by the parents just because it provides special conditions which the young may need, such as shelter, warmth, humidity or nourishment.

Animals may even be ingenious enough to use their own bodies as the best kind of home in which to raise young or nurse them through the vital early stages of life. The mouth-breeding fish provides an excellent example. As soon as the eggs of this fish are fertilized they are gobbled up, but not swallowed, by the parent and kept in the mouth for several weeks until they hatch. The young are allowed out to feed, but as soon as a predator approaches, the whole shoal shoots back into the parent's mouth, which has become their home. Many frogs use their bodies for housing their young. Darwin's frog, for example, uses its mouth, while the marsupial frog keeps its tadpoles under a special flap of skin on its back. The logical extreme of this approach is achieved by the mammals, for the developing embryos are kept inside the mother's body for long periods. Marsupials, which are primitive mammals, give birth to poorly developed young, but these immature offspring creep to a special pouch which is used as an incubator-like home for many weeks. Even in the more advanced mammals the young are adept at using their mother as a first home, clinging on to her for several months after birth.

The body or its immediate surroundings may make a convenient home, but many animals actively seek out a home or build one for themselves. Spiders, insects and crustaceans are among the many invertebrates that dig burrows. Some marine worms never leave these safety shafts; they produce their own water currents to carry in food and take away wastes.

For egg-laying animals a nest is a special sort of home which is essential to contain and protect the eggs and maintain them in the correct conditions. The labyrinth fish floats a mass of bubbles on the water surface to serve as a nest, while the stickleback sticks strands of plants together to build a nest on the bottom of a river or a pond. Most reptiles solve the egg-protection problem by burying their eggs in soft material such as sand. Turtles and alligators may hide their eggs several feet down and then leave them to hatch out. The young must struggle to the surface before making their way to the nearest stretch of water.

Of all the nest-building animals, the birds hold the prize for elaborate and painstaking constructions. Birds make use of almost every possible building material and choose a vast range of sites to ensure nest safety. Many sea birds find a sheer cliff face or a cave the safest site, while others use small islands or place their nests in marshland vegetation over water. Holes, whether natural or specially excavated, also provide extremely good protection.

Most birds' nests are located in some sort of vegetation to keep them well hidden or camouflaged. For building, the same vegetation—or other materials—are used.

Cave swiftlets construct their nests from mud and saliva, while the ovenbirds make their own clay mixture. The delicate nests of some hummingbirds are built from spiders' webs. Some of the most exotic structures of all are the hanging nests of weaver birds and the elaborate constructions of male bowerbirds. The bowerbirds' nests are really homes with ornate gardens, and their purpose is different from most homes, namely to attract the female for mating. Many nests provide valuable insulation for the eggs, but those of the mallee fowl are an attempt at more accurate temperature control. The eggs are buried in a mound of earth and decaying vegetation, and by constantly varying the depth of the covering the bird incubates them at a steady temperature until the chicks are hatched.

Mammals also build a wide variety of homes, some of them rather like birds' nests. These nests, often placed in the trees for protection from predators, are generally round and rather untidy but are larger and stronger than those of birds. The nests are often very thick, too, for mammals such as rodents use their nests for hibernation as well as for rearing their offspring. The harvest mouse is one of the few mammals capable of fine, detailed nest building. The ball-shaped nest is woven among the strong stems of cereals, and beautifully lined with fine material.

As well as building nests, mammals—particularly rodents and carnivores—are also great diggers and burrowers, and the mole remains underground for life. Many mammals use the safety and shelter afforded below ground to make extremely complicated systems of underground passages and chambers. Badger setts, for instance, may have several entrances leading into a labyrinth of passages and chambers. The sleeping and breeding areas contain heaps of bedding material brought down from the surface. Prairie dogs inhabit vast underground systems called towns, but within the town each female has her own private nesting chamber. The beaver's lodge and dam is perhaps the most remarkable mammal home and is a feat of animal engineering second to none, as it involves large-scale manipulation of the environment and a great deal of cooperation between individual animals.

It is not just the mammals that have extended the concept of a home to include alteration of the environment and the creation of large societies. Social insects such as ants and termites make some of the most ingenious and sophisticated structures in the animal kingdom. Some leaf-cutting ants construct vast underground cities where thousands of individuals live and work, while other ant species house and feed aphids and then 'milk' them for their rich, sugary secretions. Termites also live in enormous societies and have made their homes so controlled and self-sufficient that they are well protected from the harsh outside world. But the termites pay a high price for their achievement—a total dependence upon the artificial environment they create for themselves. They die if exposed to the outside for too long. Man has not yet reached such a situation, but modern city dwellers might find survival tricky if banished to the forests and plains once inhabited by their primitive ancestors.

HOME BUILDING

A place of safety is vital while the young are defenceless. It may be a nest, a township—or even the mother's mouth.

The ants' nest

No ant lives alone; they are all highly social insects that live in large colonies. Different species, however, construct vastly different nests. Some ants appear simply to make a hole in the ground, but the nest is actually quite complex. There are the army ants that do manage to exist without a nest—the workers merely form a temporary shelter, or bivouac, at night by closely surrounding the queen and her brood. The next morning they are off on the march again hunting for food.

To protect themselves from predators and the weather, many ants build nests underground. Using their mandibles as trowels to scrape away the earth, certain species dig a shallow maze of horizontal tunnels while others prefer vertical shafts with galleries leading off them. These may extend for several feet underground, depending upon the softness of the substrata, and form a labyrinth of chambers.

Of completely different design are the mound nests built by wood ants. They select particles of soil and small pieces of vegetation to construct a series of tiny chambers supported by columns and walls. The nest is continuously expanded as new stories are added. Its distinct shape causes the temperature inside the nest to rise during the day, as it catches the sun's rays from all angles. But a mound nest can lose heat at night, and so the ants insulate the outside with a layer of vegetation. At midday, larvae are often moved to the top of the mound, which is warmer, to accelerate their development. The ants constantly change the height of the mound and the amount of insulation in an attempt to regulate the temperature throughout the seasons.

Trees and plants are also chosen by ants for their nest sites. Many species bore into the softer parts of rotting trunks; others carry soil particles up into the leaves and construct their nests there. Yet others prefer the natural hollows and cavities in the stems of plants. The ants receive a ready-made home and in return the fierce soldier ants may protect the plant from being eaten by other animals. Weaver ants, *Oecophylla*, employ a more ingenious method of nest building, one which requires a great deal of cooperation from the workers. They make their nests by fastening leaves together with silk from a special gland in the ants' larvae. Keeping their hind legs on one leaf, several workers stretch out to a nearby leaf and pull it over with their front legs. Another worker then holds in its jaws a larva, which secretes silk, and, using it like a giant tube of glue, joins the two leaf surfaces with a zigzag motion.

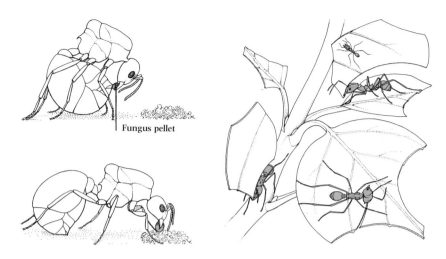

When a queen ant leaves a colony to set up her own, she takes a small pellet of fungus with her. She grows it on her own manure while laying eggs. Some 40 to 60 days later the first workers are ready.

Fungus-growing ants cut pieces of leaf and gather other plant material as a growth medium for their fungus gardens. All this material is used solely for the garden; the ants themselves eat only the fungus.

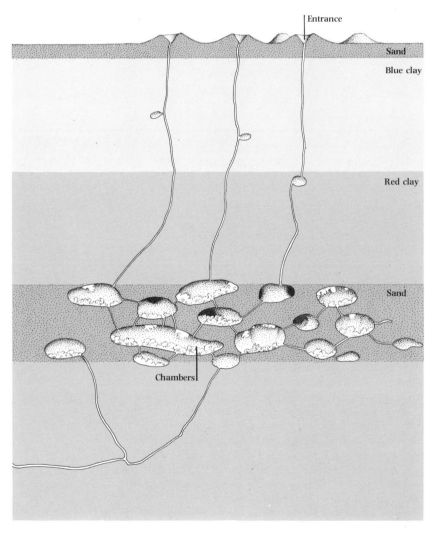

A species of leaf-cutting ant, *Atta texana*, is found in the south of the United States. It is also known as the town ant, as its vast underground nest may be 20 ft (6 m) deep and 50 ft (15 m) across. From the surface it appears as a large flat mound of soil; entrances at ground level lead down to a maze of chambers, some containing eggs and larvae, and other large ones the fungus gardens which the ants cultivate.

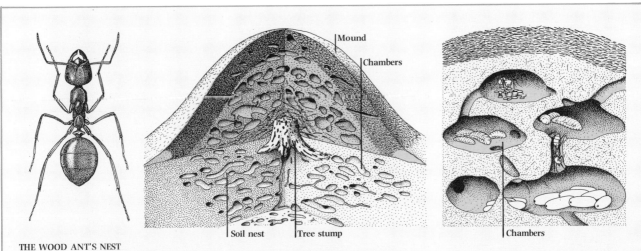

THE WOOD ANT'S NEST

Red wood ants, such as *Formica polyctena*, build imposing nest mounds several feet high. The nest is often started round a tree stump which gives support. Gradually the stump and the first passages and chambers are hidden by twigs, grass, moss, pine needles, whatever the ants can find, and the mound takes shape. A complex system of tunnels and chambers for eggs and larvae extends well underground and up into the mound itself. Materials unsuitable for building, such as sand, are removed.

A bed of fine, moist plant material provides an ideal growing medium for the fungus. When fresh leaves are brought into the nest the ants lick the leaves, cut them into tiny pieces, then chew the edges until the fragments are pulpy. They may add some anal fluid before carefully inserting the prepared leaves into the garden. The fungus grows as filaments through the bed and produces tiny swellings which the ants pluck to eat or to feed to the larvae.

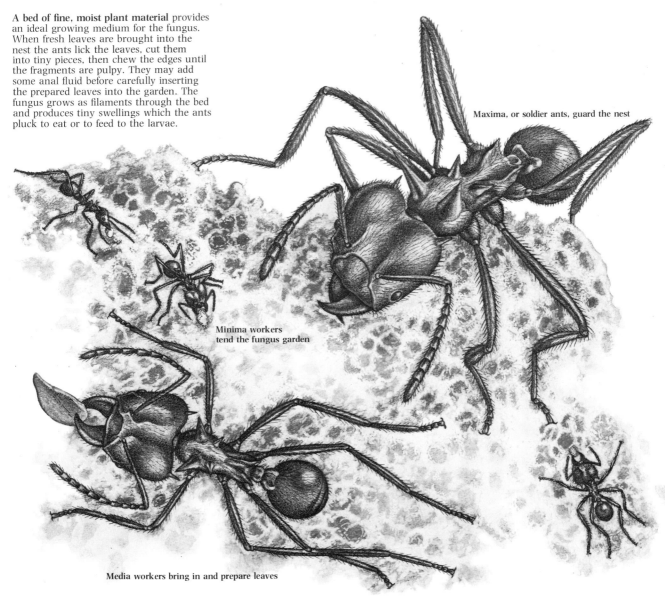

Maxima, or soldier ants, guard the nest

Minima workers tend the fungus garden

Media workers bring in and prepare leaves

The termite mound

Termites are ancient insects that perhaps pioneered true insect social life. They were probably the first insect group to solve the many problems that arise from living together in large numbers, and they also learned to manipulate their environment to protect themselves. In fact these white, soft-bodied insects have become so dependent upon their dark, warm and humid nests that they can perish within a few hours of being exposed to the open air. It is thought that their closest non-social relatives are cockroaches.

Termites live on vegetable material. Primitive species such as kalotermitids actually feed on the wood in which they nest. They digest it with the aid of protozoa which live in their guts. These species can be serious pests. More advanced termites, such as those of the family Termitidae, nest in the soil and feed on dead wood, grass and other sources of cellulose. Species of *Macrotermes* feed on fungus which they cultivate in chambers inside the nest.

In its simplest form, a wood termite nest is a number of chambers and passages in infested wood. Other termites which live in soil nests, build them by mixing faeces or soil particles with saliva. Once placed in position this building mixture hardens like cement. The visible surface mounds of the nests vary in shape: some are small and round but others, such as the mounds built by *Macrotermes* termites, are enormous hills which dominate some flat African landscapes.

The purpose of these structures is to protect the colony from predators and to provide constant and favourable climatic conditions within. Termites manage to make the walls of the nest effective insulators in order to keep in heat and moisture, as well as effective ventilators to allow oxygen to pass in and carbon dioxide to pass out. Since continual warmth and humidity are essential to the termites, the few species that live in temperate regions are active only during the warm summer months. Even those in the tropics need to build nests to ensure that heat and humidity are maintained day and night.

Termites have a highly developed social structure. The workers are immature nymphs and sterile adults of both sexes. They build nests, chambers and passages and, in the species that cultivate fungus, tend the fungus gardens. Soldiers are sterile adults with enlarged heads and jaws. They are usually larger than the workers and their job is to defend the colony. The king and queen are the reproductive individuals. In pairs they establish a colony and produce all its members.

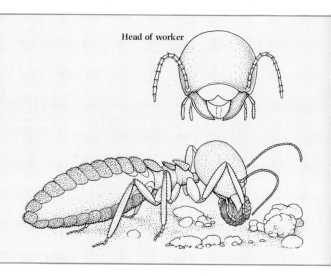

BUILDING A NEST Blind, white, wingless worker termites build the termite mound. They mix particles of sand or clay with saliva and work them together with their legs and mouthparts. The termites then press each lump firmly into place, where it sets into a hard, cementlike substance. The vertical walls of the nest are constructed first and then roofed over.

Head of worker

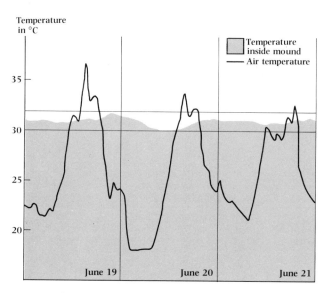

Temperature in °C

Temperature inside mound
Air temperature

June 19   June 20   June 21

Temperatures recorded inside a mound 5 ft 3 in (1.6 m) high, built by *Macrotermes bellicosus*, reveal that the temperature of the nest itself remains almost constant from day to day. Outside, the temperature fluctuates sharply between the hot days and cooler nights. The fungus chambers and the metabolisms of the thousands of insects inside the nest warm the air, and the thick wall of the mound acts as an insulating layer.

The queen termite has her own chamber where she lives with the king. Her abdomen is extremely large and in the species *Macrotermes bellicosus* she is up to 4 in (10 cm) long—16 times the size of the $\frac{1}{4}$ in (6 mm) workers. A veritable egg-laying machine, she produces thousands of eggs daily until the colony may number over a million. She is fed and attended by workers who remove the eggs to brood chambers and feed the larvae when they hatch.

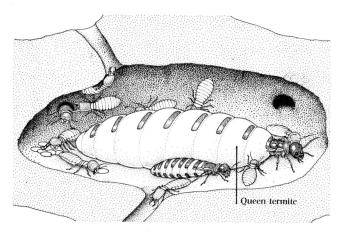

Queen termite

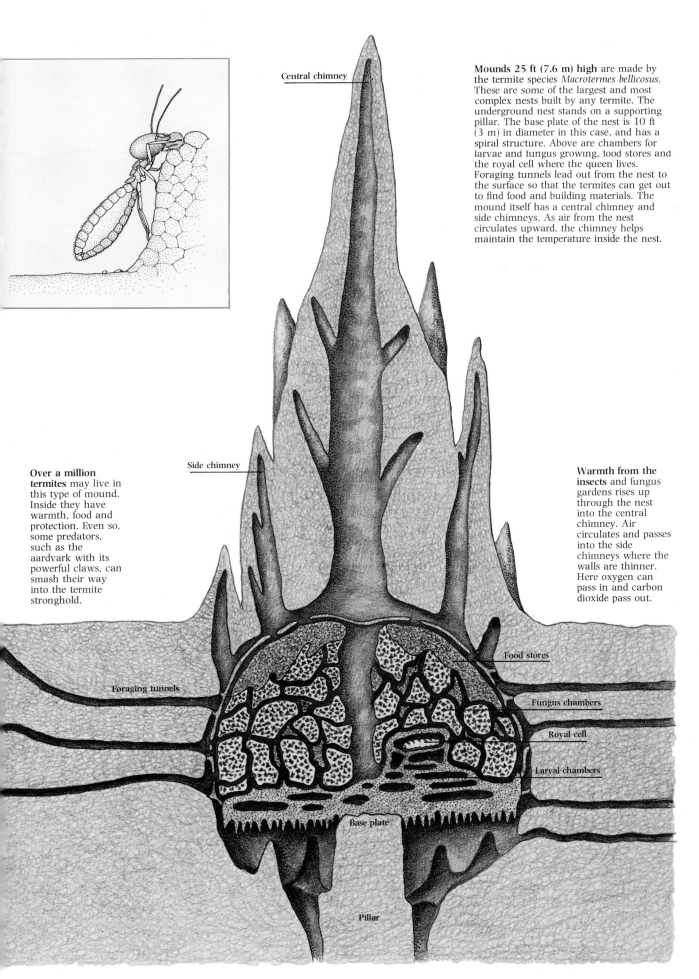

Central chimney

Mounds 25 ft (7.6 m) high are made by the termite species *Macrotermes bellicosus*. These are some of the largest and most complex nests built by any termite. The underground nest stands on a supporting pillar. The base plate of the nest is 10 ft (3 m) in diameter in this case, and has a spiral structure. Above are chambers for larvae and fungus growing, food stores and the royal cell where the queen lives. Foraging tunnels lead out from the nest to the surface so that the termites can get out to find food and building materials. The mound itself has a central chimney and side chimneys. As air from the nest circulates upward, the chimney helps maintain the temperature inside the nest.

Side chimney

Over a million termites may live in this type of mound. Inside they have warmth, food and protection. Even so, some predators, such as the aardvark with its powerful claws, can smash their way into the termite stronghold.

Warmth from the insects and fungus gardens rises up through the nest into the central chimney. Air circulates and passes into the side chimneys where the walls are thinner. Here oxygen can pass in and carbon dioxide pass out.

Foraging tunnels

Food stores

Fungus chambers

Royal cell

Larval chambers

Base plate

Pillar

A chance for survival

Parental care among the majority of fish, amphibia and reptiles is restricted to rather rudimentary protection of the eggs and the new-born young. Many fish cast their eggs at random into the upper layers of the ocean, where they are vulnerable to a host of carnivorous species. An adult codfish, *Gadus morhua*, for example, lays about two million eggs during each breeding season, from which only a tiny number will finally complete development. Similarly, amphibia deposit huge masses of eggs in strings or balls, but the survival rate of the hatched infants is extremely low. Many reptiles bury piles of eggs in soft soil, sand or rotting vegetation, but the number that survives into adulthood is also minute. With wastage running at such staggering levels the evolutionary pressure for the development of better protection is great.

Among higher animals, however, there has been a gradual evolution of nesting or protection of the young that has brought with it a reduction in waste, so that it has become normal for a large proportion of eggs to survive.

The staggering juvenile mortality among fish, amphibia and reptiles does not mean that they completely neglect their young. Basic care, which is largely instinctive, prevents a proportion of the infant deaths.

Both fish and amphibian eggs have a protective, sticky coat that enables them either to adhere to the undersides of aquatic vegetation, which are usually ignored by predatory fish, or to cling together in a huge mass, which provides those in or near the centre with a better opportunity for survival. Reptile eggs, however, are quite vulnerable. Even though the adults do not need to breed in water, they must lay their eggs in moist places to prevent dehydration. Since the leathery coat surrounding each egg must be thin enough to enable the developing embryo to breathe, it cannot be sufficiently tough to give complete protection from predators.

Some fish and reptiles do make practical provisions for their young. Turtles labour up the beach to lay their eggs in pits they dig in the sand. The male stickleback actually constructs a nest from bits of alga glued together with his own secretions.

THE GREEN TURTLE
After mating at sea the female turtle travels up the beach and excavates a pit about 15 in (38 cm) deep in which she lies, her shell flush with the beach. She then digs another hole with her rear flippers, usually about 16 in (40 cm) deep, for the eggs. She lays about 100 eggs, each about 2 in (5 cm) in diameter. She may lay more clutches over the next few weeks. The eggs hatch within 50 to 80 days depending on the latitude. The young hatchlings then have to scramble out of the nest. Mass effort is required to dig themselves to the surface—another advantage of the large number of eggs. Of an estimated 1,800 eggs laid during a female's lifetime, only a few survive.

With its front flippers the turtle scrapes a body pit in the sand. Then she employs her rear flippers with a downward scooping action to dig the egg pit. The pit can be as deep as the reach of the rear flippers.

The labyrinth fish, *Macropodus* spp., lives in slow-moving stagnant waters, where it survives because of its ability to breathe air. At spawning time the male snaps for air more frequently, creating bubbles on the surface which are strengthened by a special saliva. When the female lays her eggs they contain a buoyant oil droplet which carries them up to the surface where they lodge in the bubble nest. The male guards the eggs.

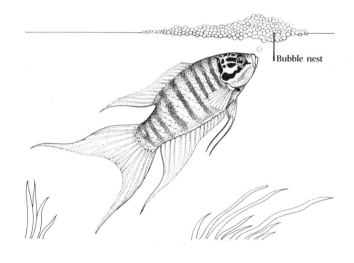

Bubble nest

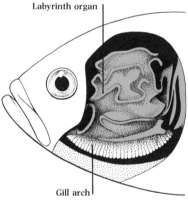

Labyrinth organ

Gill arch

Their ability to gulp in oxygen from the air enables labyrinth fish to live in water poor in oxygen. They have an accessory breathing organ, the labyrinth, consisting of a much folded vascular membrane through which oxygen can be extracted.

A degree of parental care is shown by some snakes, such as the primitive pythons and boas (family Boidae). After laying a clutch of up to 100 eggs, the female snake winds herself round them and remains on guard. By a gradual shuffling process she can move them to catch the early morning sun or seek the midday shade. After they have hatched she pays the same attention to the baby snakes.

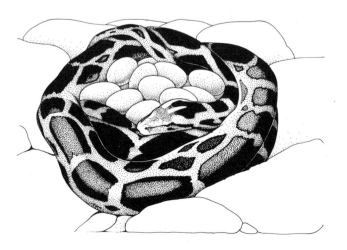

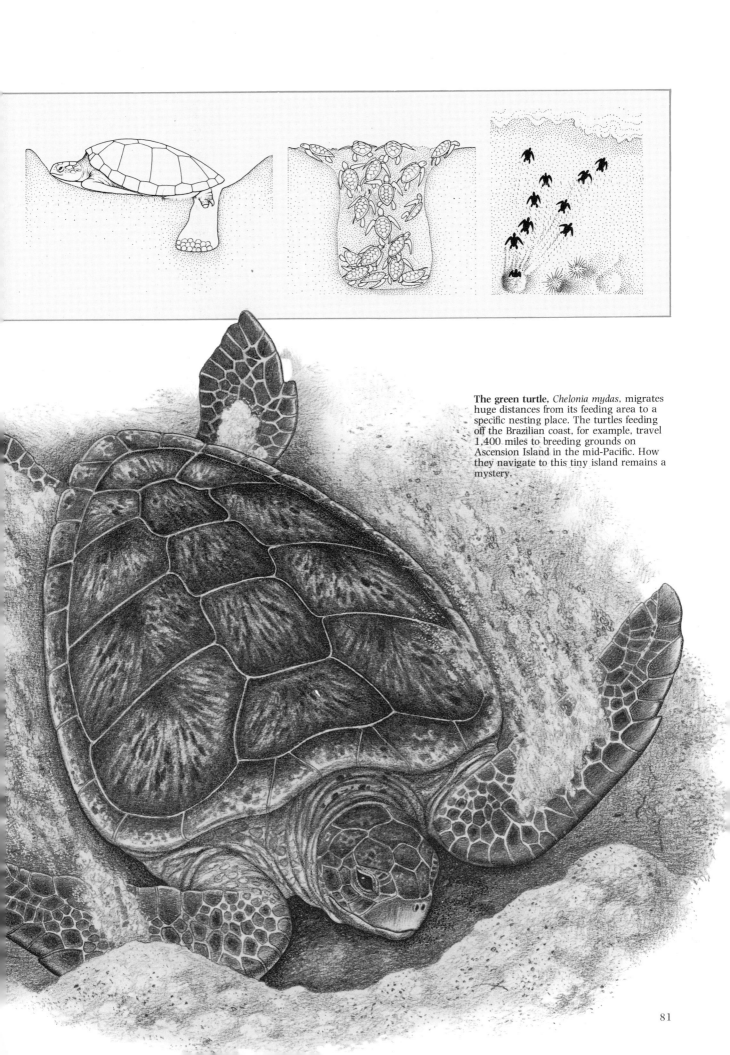

The green turtle, *Chelonia mydas*, migrates huge distances from its feeding area to a specific nesting place. The turtles feeding off the Brazilian coast, for example, travel 1,400 miles to breeding grounds on Ascension Island in the mid-Pacific. How they navigate to this tiny island remains a mystery.

The body as a home

Instead of building a nest to protect their young, many animals keep their infants with them until they are old enough to fend for themselves. One of the simplest arrangements is seen in monkeys and bats; the infant just clings tightly to its mother's fur. In these species, the young are born with well-developed hands and feet, and the clinging response seems to be innate. Another basic adaptation is found in species that usually give birth to twins, such as the marmosets, family Callitrichidae. As a rule, the mother is too small to transport both infants on her back. Fortunately, the bond is so strong between the parents that the male stays with the female long after the young are born and carries one of the babies.

A slightly more advanced form of 'piggyback' transport is exhibited by the pangolins, or scaly anteaters, order Pholidota, of Africa and India. As the female enters puberty, she develops scales at the base of the spine that have specially ridged edges. These give the new-born pangolins a convenient handle for clinging to their mother's otherwise slippery armour-plated back.

Although shrews give birth to their young in secure nests, the female is unable to provide enough food for them as they approach the weaning period. Since there are normally six youngsters in a litter, the mother cannot carry all of them on her back. Consequently, when it is time to forage for food, each holds on to the tail of the one in front by its teeth, with the foremost similarly linked to the mother, thus forming a caravan.

Many species have extreme physical adaptations for protecting their young. Midwife toads, *Alytes*, for example, carry a bundle of spawn on their legs, which they shake off into a nearby pool when the eggs are ready to hatch into tad-poles. Other toads spread their eggs over their backs, then grow a flap of skin to cover them. The eggs stay there in complete safety until they are ready to emerge as tiny toads. Some sharks, lizards and snakes also keep their eggs inside them until they hatch. But these eggs do not continuously derive their nourishment from the mother, though they do get their water from her, and the process is quite different from the mammalian placenta system.

Another highly specialized adaptation is seen in the mouth-breeder fish, family Cichlidae, which inhabits several African lakes. It carries its eggs, and, later when they have hatched, its fry, in its mouth, letting them out to feed only when there is no danger of predators. The adults feed on tiny creatures no bigger than the baby fish and it is a mystery how they know not to swallow their own young.

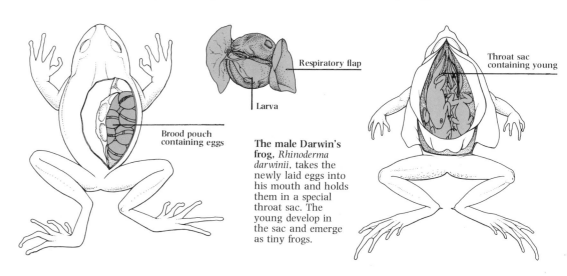

As soon as the female lays her eggs, the male marsupial frog, *Gastrotheca marsupiata*, sweeps them up and smears them on his back with his hind feet. The eggs are sticky and do not roll off. Within a few hours, the skin on his back softens and the eggs are drawn into tiny ponds. A sheet of skin grows over the ponds in his back. The larvae complete their development on the father's back.

Brood pouch containing eggs

Respiratory flap

Larva

The male Darwin's frog, *Rhinoderma darwinii*, takes the newly laid eggs into his mouth and holds them in a special throat sac. The young develop in the sac and emerge as tiny frogs.

Throat sac containing young

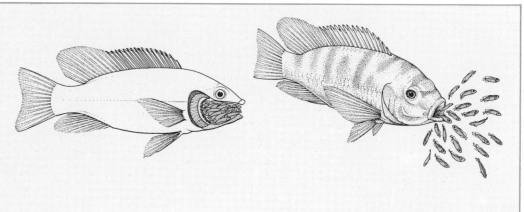

MOUTH-BROODING

The male or female mouth-breeder fish, *Tilapia* sp., snaps up the eggs immediately they are laid. For the next 5 weeks the parent keeps the developing young in its mouth. As the young grow, the parent lets them out for foraging, but if danger threatens the shoal returns to the adult's mouth.

Many, but not all, marsupials carry their young about in a pouch on the body. The young marsupials are born after a gestation of as little as 11.4 days (the bandicoot) in a very undeveloped state and continue development inside the pouch. As they grow they spend more and more time outside the pouch but always return to it for milk, for protection and for warmth.

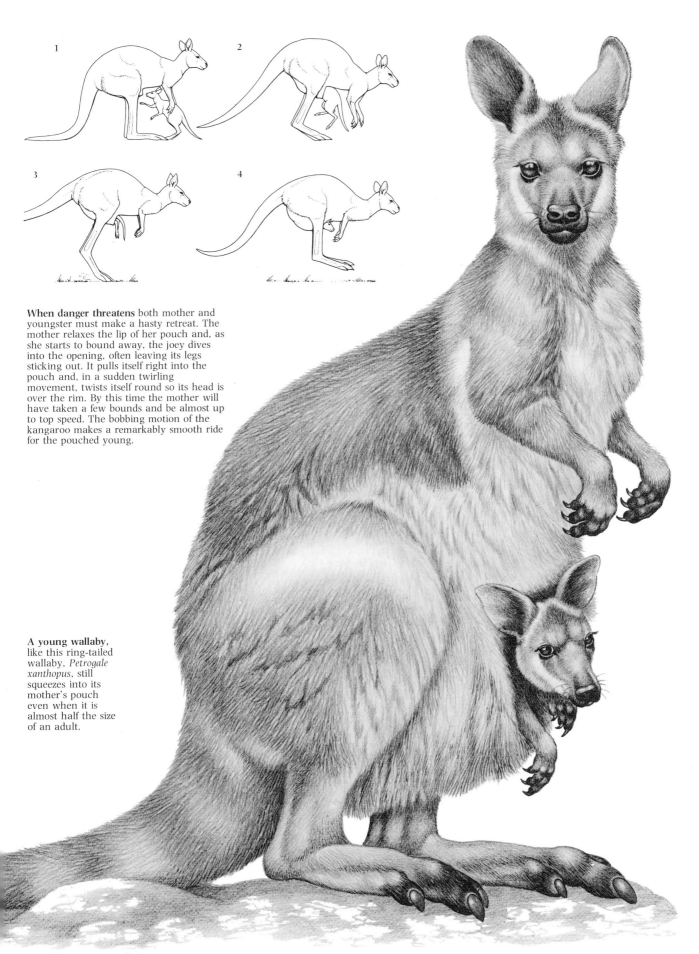

1 2 3 4

When danger threatens both mother and youngster must make a hasty retreat. The mother relaxes the lip of her pouch and, as she starts to bound away, the joey dives into the opening, often leaving its legs sticking out. It pulls itself right into the pouch and, in a sudden twirling movement, twists itself round so its head is over the rim. By this time the mother will have taken a few bounds and be almost up to top speed. The bobbing motion of the kangaroo makes a remarkably smooth ride for the pouched young.

A young wallaby, like this ring-tailed wallaby, *Petrogale xanthopus*, still squeezes into its mother's pouch even when it is almost half the size of an adult.

The way of the birds

Keeping eggs and young safe from predators is a problem that birds solve in many different ways. There are a few species which rely on well-concealed places rather than nests. Sea birds, such as waders and gulls, merely scrape out a shallow depression in the sand or shingle above the hightide mark. Since the parents do not cover the eggs with nesting material, they take turns sitting on and incubating them. If both parents have to leave the nest, the eggs are exposed not only to cold air and rain, but also to predators.

Other sea birds protect their eggs by nesting in a virtually inaccessible place. They often choose a rocky ledge on a sheer cliff face to lay their large, white, conspicuous eggs. The pointed oval shape of the eggs prevents them from rolling off the cliff. Moreover, since most sea birds live in huge colonies, they rely on sheer strength of numbers to drive off any predators. Still other bird species have found that a nest site on or near water will deter predators. Some grebes and coots construct a floating raft of vegetation for their nests, and reed warblers weave their nests high in waterside plants.

On dry land solitary birds that are small and camouflaged build their nests in well-hidden places, far away from others. This minimizes their chances of being found by predators. Studies have shown that once a predator finds a nest in a particular situation, it will immediately look nearby for others.

Trees are favourite nesting sites for many birds. Some species lodge their nests in the forks of trees, or suspend them from the tips of branches because they are difficult to reach there. Holes in trees provide the safest nests, especially if the hole is small enough to keep out large predators. Species such as woodpeckers are capable of making their own holes, but many others make use of old holes or those produced naturally by damage or rotting. Holes in cliffs or on the ground are also popular nesting sites; it is not unusual for the latter to have been dug originally by a rabbit or a rodent. There are even birds that occupy the abandoned nests of other bird species—falcons are known to take over crows' nests.

Depending upon their complexity, nests may serve several functions. Besides holding and concealing the eggs, they also keep them warm by providing insulation against heat loss. In general, females build the nests, but there are exceptions. Male wrens, for example, make several nests in their territories before the females arrive. In this case, the nests may have another purpose— that of attracting females.

Emperor penguin,
Aptenodytes forsteri

Little tern,
Sterna albifrons

Skylark,
Alauda arvensis

2¾ in (7 cm) across
inside of nest cup

Coot,
Fulica atra

Up
(30

Burrowing owl,
Speotyto cunicularia

Burrow 18 in (45 cm)
to 6 ft (1.8 m) long

Manx shearwater,
Puffinus puffinus

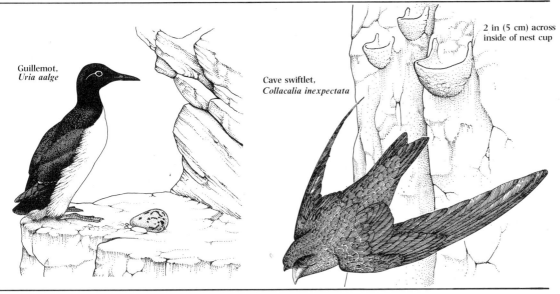

Guillemot,
Uria aalge

Cave swiftlet,
Collacalia inexpectata

2 in (5 cm) across
inside of nest cup

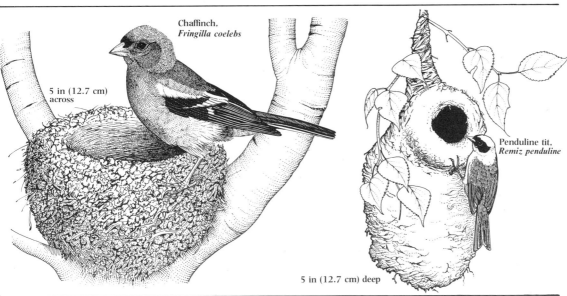

Chaffinch,
Fringilla coelebs

5 in (12.7 cm)
across

Penduline tit,
Remiz penduline

5 in (12.7 cm) deep

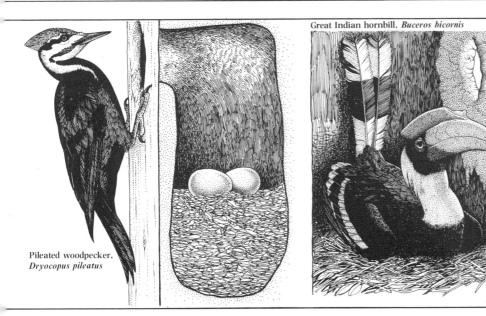

Great Indian hornbill, *Buceros bicornis*

Pileated woodpecker,
Dryocopus pileatus

Some penguins, such as the emperor and king penguins, manage without nests. They incubate a single egg on their feet, covering it with a fold of skin. Terns make a rudimentary scrape on the shore and rely on camouflage, and the colony, to protect the eggs. A nest on a sheer cliff face is safe from most predators. Sea birds, such as guillemots, nest on narrow ledges hundreds of feet above the open sea. Swallows and swifts make mud nests on cliffs and walls. Cave swiftlets produce a gelatinous saliva to construct their nests.

Ground level nests are at great risk from predators. If a nest must be on open ground, it should be well camouflaged in vegetation, as is the skylark's nest. Another means of protection is to build a nest platform in shallow water—a technique used by waterside birds like the coot. Nests in bushes and trees may keep eggs out of the reach of predators. The finch selects a convenient fork to hide and support its nest. The penduline tit suspends its more obvious nest from the end of a branch.

Holes in the ground are an effective means of protection. The burrowing owl can dig its own burrow, but usually occupies one vacated by prairie dogs. Old rabbit burrows on sandy cliffs are a common nesting site of manx shearwaters. One of the safest of all nest sites is a hole in a tree. Woodpeckers can make their own tree holes but may rely on natural ones. Once inside the tree with her eggs the female hornbill walls herself in using mud brought to her by her mate. He then feeds her during her confinement.

Nesting techniques

Remarkably, a young bird can build a perfect nest the first time it tries. The expertise for this complicated activity therefore appears to be largely inherited. Although there are a few notable exceptions, in most species it is the female that undertakes the task of nest building.

Since each species builds its own kind of nest fitted for particular ecological needs, it follows that nests vary in shape and construction. The simplest ones are found among many of the sea birds, whose rudimentary containers are merely piles of twigs. Some of the large birds of prey, such as eagles, build enormous untidy heaps of sticks at the tops of tall trees.

The most intricate nests, however, are constructed by smaller birds. They use delicate vegetation, such as grasses, which they can press and even weave into place with their small bills. Sometimes they dip the vegetation into water to make it more pliable, as well as tighter and stronger after drying. Basically, a bird makes the familiar cup-shape by firmly pressing its body into the nest. In the later stages, it may sit in the cup while finishing off the outside work to ensure a good fit. The cup itself is often lined with soft materials, such as moss, grass, old feathers and horse hair.

Many nests are enclosed with a domed top and have a side entrance or tunnel. Weaver birds, for example, make a variety of elaborate nest shapes; some are round with an entrance near the top; others resemble a long sock with an entrance underneath. Hummingbirds build the most delicate nests of all. At the most an inch across, their miniature structures are made of plant down but occasionally include gossamer from spiders' webs.

Certain birds, however, have abandoned vegetation completely. Swifts and swallows seek out damp mud or clay near pools to construct hard nests on cliffs or buildings. Mud is also utilized by some species to reinforce a twig nest or line it. The ovenbird builds a nest of wet clay which then bakes hard in the sun.

The beak is the principal, and for many birds the only, tool that is available for nest building. In species such as the woodpecker, the bill is chisel-shaped and thus ideal for drilling a nest hole. Among the burrowing birds, a number use their powerful feet, as well as their beaks to excavate holes in sandy soil for their eggs. Feet are also a useful tool for the birds that weave nests.

THE KINGFISHER'S BURROW
The kingfisher is one of the few birds to excavate its own burrow for nesting. Kingfishers usually have a number of favourite perches over a stream, and from one of these they select a nesting site. The soil of the river bank must be soft enough to dig into, yet firm enough to support a tunnel. The bird dives at the river bank to make the initial inlet. It continues until the hole is big enough for it to land and get a grip. As it digs deeper, the kingfisher uses its feet too, flinging out soil with strong backward movements. A deep safe chamber, with enough room for the bird to move around, is made at the end of the tunnel, and there the eggs are laid.

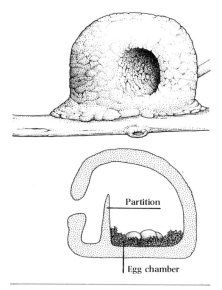

Both partners help build the nest. They wait for rain to soften the clay soil and then collect mud in their beaks. By pressing the lumps of clay together with their beaks and feet they construct the nest. Over 2,000 pieces of clay may be gathered in all. The hot sun soon bakes the clay into its final ovenlike shape. A curved partition separates the entrance from the brooding area, allowing just enough room for the birds to get in, but keeping large predators out.

Partition

Egg chamber

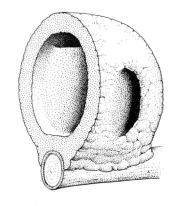

Ovenbirds, such as *Furnarius rufus*, build remarkable clay nests, usually on the branches of trees. These true ovenbirds are of the family Furnariidae and live in Central and tropical South America. The North American ovenbird, *Seiurus* *aurocapillus*, is from a quite different family, the wood warblers. The ovenbird's eggs are laid in a grass-lined inner chamber of the nest. Once hatched, the fledglings do not stay in the nest for long, as the summer sun soon makes it unbearably hot.

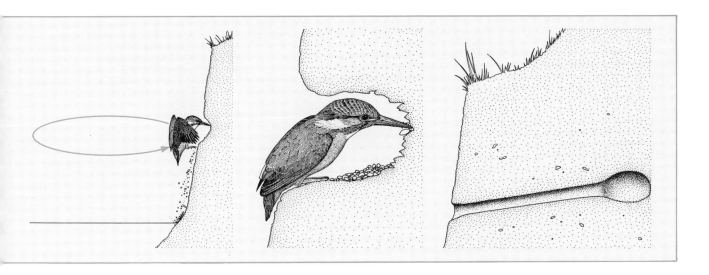

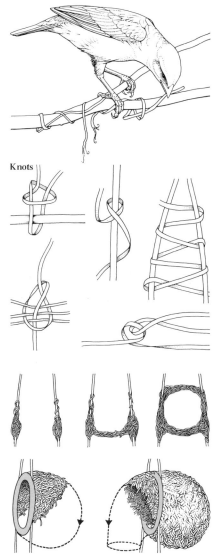

Knots

By knotting, twining and weaving with its feet and beak, the weaver bird makes its nest. It gathers long pieces of grass or palm fronds and uses them to knot twigs together for a framework. More material is added to make a circle from which the nest hangs.

Weaver birds make nests which are not only marvels of construction but also extremely difficult for a predator to approach and enter. Typically, the suspended nests are woven from vegetable fibres, but there are many shapes. Some,

like the nest of the spotted-back or village weaver, *Ploceus cucullatus*, above, are kidney-shaped with an entrance at the bottom. Others are sock-shaped, again with a bottom entrance, or rounded with an entrance at the side. Some social weavers

build a platform and from it suspend a number of nests. Generally the male bird makes the nest, and this may take him a whole day. Once he has attracted a mate she may help with the inner lining. Most weaver birds live in Africa.

Incubators and bowers

Most nests are built by birds to house their eggs and young. But in some species, the megapodes of Australia and the bowerbirds of Australia and New Guinea, the nest has special functions.

Megapodes are secretive, ground-dwelling birds that rarely fly. Although found in various habitats, they all construct nesting mounds in which they bury their eggs. Some species dig pits in the sand and leave their eggs to hatch. Others take advantage of hot volcanic soil to help incubation. One of the most remarkable is the junglefowl, *Megapodius freycinet*, which constructs huge mounds of earth up to 35 feet (10.6 metres) wide and 15 feet (4.5 metres) high. These contain vegetable matter which decomposes and so raises the temperature inside the mound, thus helping to incubate the eggs. Another megapode, the mallee fowl, also builds incubator mounds containing vegetation, and the male regularly tests the temperature by probing the mound with his bill and tongue. If it is too hot, he will remove some material and even expose the eggs which are normally deep in the mound. Since the eggs are laid over several months, the male spends most of the year as a diligent gardener, testing and attending to the mound. When the young hatch, they dig themselves out and quickly vanish into the surrounding area without even seeing their parents.

Bowerbirds, on the other hand, build a variety of elaborate structures (bowers) on the ground to attract females. After mating, the female lays her eggs in a separate nest made in a tree, often some distance from the bower, and incubates them alone. Even the simple bowers, which are only heaps or platforms, always contain a mixture of bright, attractive decorations. Some species select a growing sapling for their bower and decorate it like a maypole; this may reach a height of 9 feet (2.7 metres). Others construct a hut, sometimes fronted by a stockade, and adorn it with coloured objects, such as fruit, flowers or dead insects. Any brilliantly coloured human property may also be chosen to enhance the display. There are also species that construct a bower with parallel walls of twigs that may arch over to form a tunnel. The male then adds decorations to the walls or to select areas in the middle or at the ends. He may even paint the inner walls with fruit pulp or charcoal. During the display, the male picks up a few of the bright objects and waves them in front of the female.

Because females do the choosing and tend to select males with the most attractive bowers, competition among the birds is intense. But as the male plays no part in rearing the young, he can afford to expend his energy in devising bigger and more ornate bowers with which to lure females.

The mallee fowl, *Leipoa ocellata*, lives in arid, inland areas of Australia. Instead of sitting on its eggs it incubates them in a mound. The male bird digs a pit in the ground up to 15 ft (4.5 m) across and 4 ft (1.2 m) deep. During the winter he fills it with plant material and leaf mould. When this layer has been moistened by the rain he covers the pit with a layer of sandy soil about 2 ft (60 cm) thick. Sealed off in this way the vegetation decomposes and becomes warm. By varying the thickness of the top layer of soil, he keeps the eggs at around 92 degrees Fahrenheit.

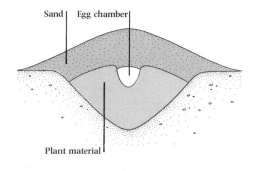

Sand | Egg chamber | Plant material

To attract a female, the male bowerbird may spend hours tending and arranging his ornamental garden and bower. He often dances to impress the female further. After mating she leaves him and lays her eggs in her own nest. She rears the young on her own.

When the female is ready to lay her eggs the male mallee fowl must first dig a hole in the mound for her. She lays her eggs in this hole one at a time at intervals of several days. The final clutch may be anything from 5 to 35 eggs. When a chick hatches, it must struggle unaided to the surface of the mound. At first after this journey it is weak, but in an hour the chick can run, and within a day it can fly proficiently. Parental care is confined to tending the incubating eggs in the mound.

The striped gardener bowerbird, *Amblyornis subalaris*, lives in the mountains of New Guinea. The male bird is about 9 in (23 cm) long and has olive-brown plumage with a crest of orange feathers. His bower is a dome-shaped mass of twigs about 2 ft (60 cm) high with openings at the front. It has a central column of a sapling covered with interlocking twigs and moss. Between this column and the outer wall of twigs is a passageway which forms a tunnel at the back of the bower. The open area in front of the bower has a dark base of tree fern fibre. On this the bowerbird places a variety of decoration such as flowers, berries, fruits and even shiny beetles in the hope of catching a female's eye.

Central column

89

The beaver dam

To provide a safe home for their families, many medium-sized mammals build their nests out of sticks and twigs. The most complex of these structures is made by the beaver, *Castor canadensis*.

An animal similar to the beaver in its way of life is the muskrat, *Ondatra zibethica*. It chooses swampy places in North America to build its globular lodges of sticks plastered with mud. Nests of sticks are also made by desert rodent species in North America and Australia, primarily to help them withstand the heat of the midday sun. The Australian stick rat, *Leporillus conditor*, interweaves sticks between 6 and 12 inches (15 to 30 centimetres) long around a bush until it constructs a pile about 3 feet (90 centimetres) high. Sometimes the nest is built over a rabbit burrow, which the rat also uses. If there are any stones the size of tennis balls available, it will incorporate these into the fabric of the nest. There are usually two or three entrances, any of which the rat may close by pulling a stick behind it into the nest.

The beaver has long been a source of interest to man, mainly because of its fabled engineering ability. By constructing a dam of sticks and mud, beavers can make small streams into wide, shallow lakes over a period of several years. These, in turn, become wider as the beavers continually fell the trees at the water's edge. The flow of water slows down so that suspended silt in the river water is gradually deposited in the lake. As the silting process continues, the lake begins to support aquatic vegetation. When this eventually dies, a peat and humus layer develops that encourages the spread of grasses and other terrestrial plants. By the time the beavers have left an area, it is a damp meadow—a beaver meadow. In fact, part of Montreal is built upon beaver meadows, and some of the best agricultural land of the central and northern United States owes its richness to the silt previously deposited there by the activities of beavers.

In the autumn, beavers undertake the long job of repairing and rebuilding their homes and dams. But in the summer they leave their lakes, dams and lodges to travel up-stream to find fresh shoots and buds. This is when they are most vulnerable to human exploitation. They have long been trapped for their fur, meat, castoreum (a secretion from their anal glands said to have medicinal properties) and front teeth, which are so sharp that American Indians and ancient Europeans used them as knives. As a result, beavers in western Europe are almost extinct. In the United States their numbers are once again increasing after they were almost exterminated.

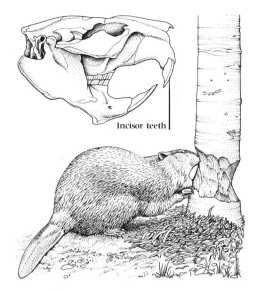

The beaver's skull is distinguished by the massive incisor teeth. They are deep orange in colour because of the hard enamel which coats their outer surfaces. As the upper and lower incisors bite on one another, the softer, inner surface enamel and central dentine wears down faster than the hard, outer coating. This is left projecting as a sharp cutting edge. A large beaver can fell a tree 8 in (20 cm) across in minutes using the biting and chipping actions of its strong teeth.

Incisor teeth

The task of building and repairing the dam is never finished. Mud, leaves, twigs and even grass are constantly incorporated, and because of the progressive silting-up of the lake the dams need regular heightening until they may be several feet high.

The dam also makes a larder for winter food when the lake is frozen. The beavers cut extra branches in the autumn and store them under the water near the dam. Thus they have fresh food right through the winter.

Mud

Tunnel

Beavers have a superb engineering ability. When the lakeside trees have all been felled for use in the dam they dig canals into the forest to bring down trees cut farther afield. To protect themselves from predators they plaster the sides of the lodge with mud, which freezes hard in the winter, but leave a ventilation shaft at the

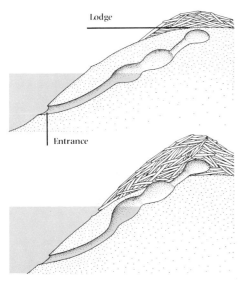

Lodge

Entrance

A beaver lodge starts with the beavers digging a burrow into the bank of a stream. The foundations of the dam are laid, and soon the entrance to the burrow, originally just below the water surface, is well below it. Sticks of various sizes are piled on the bank and woven into a circular corral, which then becomes part of the beavers' lodge. The height of the dam is adjusted so that the base of the lodge is covered by about 1½ to 2 ft (46 to 60 cm) of water.

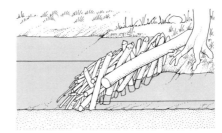

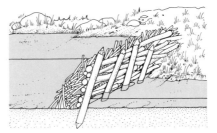

The structure of the beaver dam depends on the specific characteristics of the stream. Sometimes a tree is felled and braced against another, still growing, tree. Short, stout timbers are laid against the brace, and sticks and twigs used to plug the gaps. Alternatively, stout stakes are buried in the river bottom and stones used to wedge other sticks along the base of the row of stakes. The dam is always built of heavy timbers first, then smaller sticks.

Beavers are well adapted for an aquatic life. Their ears and nostrils close off when they are under water and they can remain submerged for up to 15 minutes.

The lodge floor is used for sleeping and for feeding the young. It is kept clean; wood chips and other rubbish are removed and fresh mud is brought in to repair and extend it.

Young beavers are born in the spring and spend a couple of months in the lodge before venturing out. They are sexually mature in 3 to 4 years.

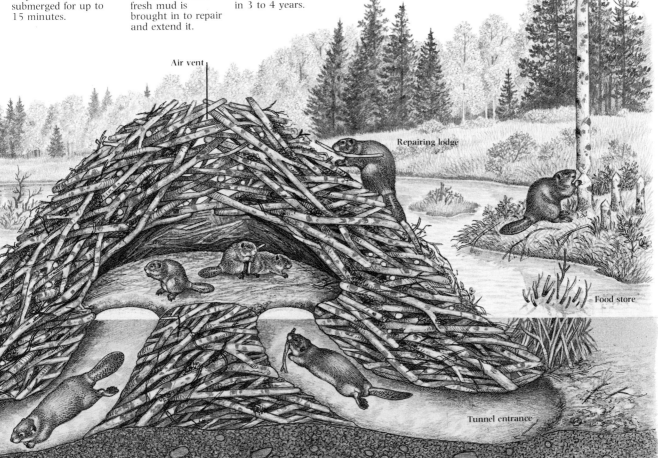

Air vent

Repairing lodge

Food store

Tunnel entrance

top. A beaver is usually up to 4 ft 3 in (130 cm) long including the tail, and weighs up to 66 lb (30 kg). The forefeet have strong claws for digging and carrying and the hind feet are webbed. The dense fur provides waterproofing and insulation. Beavers are outstanding swimmers, using their long flat tail as a rudder. They are herbivorous, feeding on shrubs and buds from trees such as poplars, aspens and willows in summer, and bark from trees and shrubs in winter.

Living underground

The burrowing habit is common among mammals since the majority of these animals live either immediately beneath or on the surface of the earth. Although largely a feature of mammalian life, burrows are also dug by snakes, lizards, turtles, certain frogs and even some fish and invertebrates to give their developing young protection. Most birds, however, build intricate nests in trees and bushes or nest high on cliffs or other inaccessible places. A few, such as the Manx shearwater, *Puffinus puffinus*, the puffin, *Fratercula arctica*, and the shelduck, *Tadorna tadorna*, nest in disused rabbit burrows, but they do little digging themselves. They use their feet to scrape away dead leaves and loose earth, as well as their beaks to remove large pieces of soil.

Burrowing mammals usually dig their own tunnels and galleries, although some, like the burrow-nesting birds, take over the abandoned burrows of others. The Cape hunting dog, *Lycaon pictus*, is normally nomadic, but during the breeding season it seeks out the disused burrow of an aardvark, *Orycteropus afer*, in which to bear its young and look after them until they are weaned. The hunting dog carries out minor reconstruction of the burrow, and lines the nest chamber with soft hair from her belly.

A wide range of special burrowing adaptations is found in mammals. Some species spend their entire lives below ground, never venturing to the surface. The most common of these is the mole, which tunnels for hundreds of feet without stopping. Its broad front feet, which have an extra 'finger', the sesamoid bone, are shaped like shovels. The mole has fine, soft fur, but the mole rat, a rodent specialized for life below the surface, is naked. Its skin is tough, wrinkled and resistant to the buffeting it receives in the tunnels. The mole rat also has slightly broadened feet, but its main digging devices are its teeth, which work at the soil like a coal miner's pickaxe. Both moles and mole rats are virtually blind, and use touch and smell for communication. Most burrowing species, however, do not show such drastic adaptations.

As a rule, small rodents breed in a special nesting burrow, which is often separate from the living burrow system of the colony. These breeding burrows are usually only blind pockets dug into the surface with one entrance. Voles and mice plug this entrance with a wad of grass and soil so that passing predators will not notice the burrow. After the young are weaned, the breeding chamber is abandoned and the young family moves in with the rest of the colony.

THE BADGER SETT
Badgers spend a considerable time gathering fresh bedding for their underground chambers. In summer they may use hay, in winter, dried ferns, and in spring, leaves of woodland flowers. To gather bedding, the badger scrapes up the pieces and moulds them into a controllable bundle under its chin. Then it moves backward, almost sliding on its forelimbs, to the sett entrance and drags the bundle down into the tunnel.

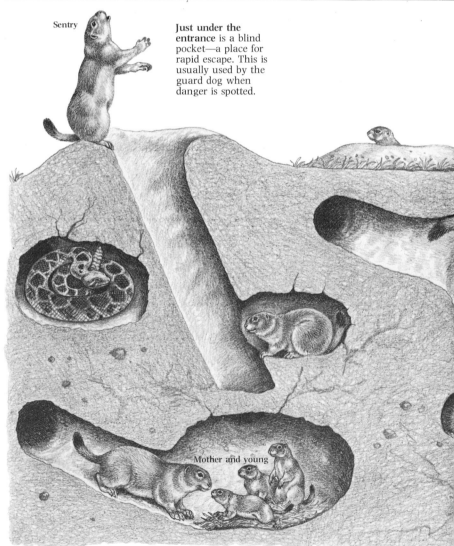

Sentry

Just under the entrance is a blind pocket—a place for rapid escape. This is usually used by the guard dog when danger is spotted.

Mother and young

Adult female prairie dogs each have their own nest chamber in which to give birth to their young. This is fiercely defended. The young spend the first few weeks of their life in this chamber feeding on their mother's milk. A little dry seed food is kept in the nest chamber and is probably used as the youngsters are weaned. As the

The number of badgers living in a sett is often between 5 and 12. From the entrance a short, straight tunnel extends into the hill for up to 6.5 ft (2 m) then turns and descends about 10 ft (3 m). There are many chambers, some for breeding and some for sleeping. Excavated soil is piled up outside the sett but some is left just inside the entrance to restrict its size. Badgers are strong diggers and can remove boulders weighing 9 lb (4 kg).

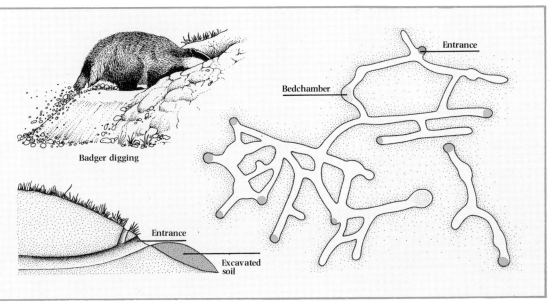

Badger digging

Entrance

Bedchamber

Entrance

Excavated soil

The prairie dog, *Cynomys ludovicianus*, is one of the most common grassland rodents in the United States. Their populations are divided into clearly defined units of coteries, wards and towns.

A coterie consists of an adult male, 2 to 4 adult females and a number of young. Several coteries make a ward and there are 2 or more wards. A town, as the whole group is called, may cover several acres.

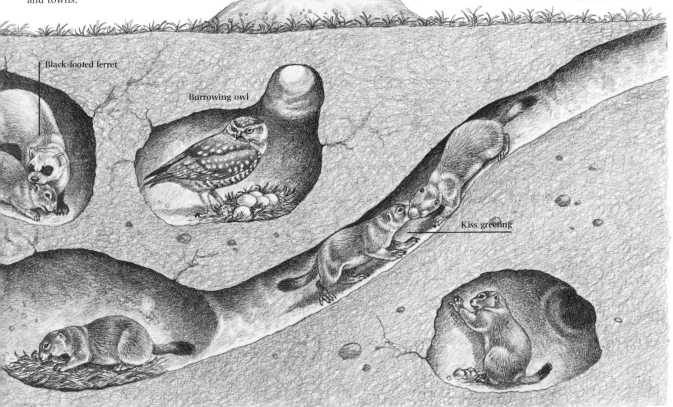

Black-footed ferret

Burrowing owl

Kiss greeting

young grow they spend more time playing in the galleries and on the surface above the burrow, always under the watchful eye of a guard. But even beneath the surface

they are not completely safe from predators. The burrowing owl, *Speotyto cunicularia*, will sometimes take over a disused burrow in the colony and prey on

the prairie dogs. Black-footed ferrets, *Putorius nigripes*, enter the burrows and against these powerful attackers the prairie dogs have no defence.

Nests above ground

The aerial world belongs chiefly to the birds, but some species of mammals have adapted to a life above ground. Some, such as the harvest mouse and the tree squirrel, spend part of their lives on the ground, but return aloft primarily for breeding. Others, such as the aerial marsupial Leadbeater's possum, *Gymnobelideus leadbeateri*, and the many monkeys, spend their entire lives high among the treetops. To bring up their young they either construct nests similar to those of birds or they utilize the hollow trunks of old trees. If they use existing tree holes, they make a lining which serves them as a soft bed.

Most tree-living mammals exhibit specific nesting behaviour and use particular materials, but an exception is the common Australian ring-tailed possum, *Pseudocheirus peregrinus*. Typically, this possum builds an untidy nest of twigs and leaves, measuring about 12 inches (30 centimetres) across, usually in thick ti tree bushes. In suburban back gardens, however, it sometimes constructs open-topped nests, resembling those of blackbirds, and lines them with feathers, wool or other foreign matter it can find nearby.

Since arboreal mammals are heavier than tree-nesting birds, their nests are generally more robust. Tree squirrels, for example, build nests called dreys from thick twigs that they wedge between the growing branches of their chosen tree. The walls may be up to 4 inches (10 centimetres) thick and the entrance is located in a position difficult for predators to find. The dreys are lined with fur plucked from the female's belly, or with finely chopped leaves, bark or grass.

Hole-nesting species, such as the marsupial sugar glider, *Petaurus breviceps*, the marsupial greater glider, *Schoinobates volans*, and the North American flying squirrel, *Glaucomys volans*, select a deep hole with a narrow entrance for their nests. Although each female flying squirrel chooses a separate nest for the birth of her own litter, during the winter as many as a dozen squirrels may be found huddled together to keep warm. The majority of arboreal marsupials line their breeding nests with chopped bark; the rodents prefer to use leaves.

The most exquisite mammalian aerial nest is built by the harvest mouse, *Micromys minutus*, of Europe and *Reithrodontomys* of the United States. Both species weave a globular nest that will neither shake nor fall apart, suspended among growing grass stems. Many other small rodents, such as the dormouse, make nests above ground, but these are untidy like the nests of the larger tree squirrels.

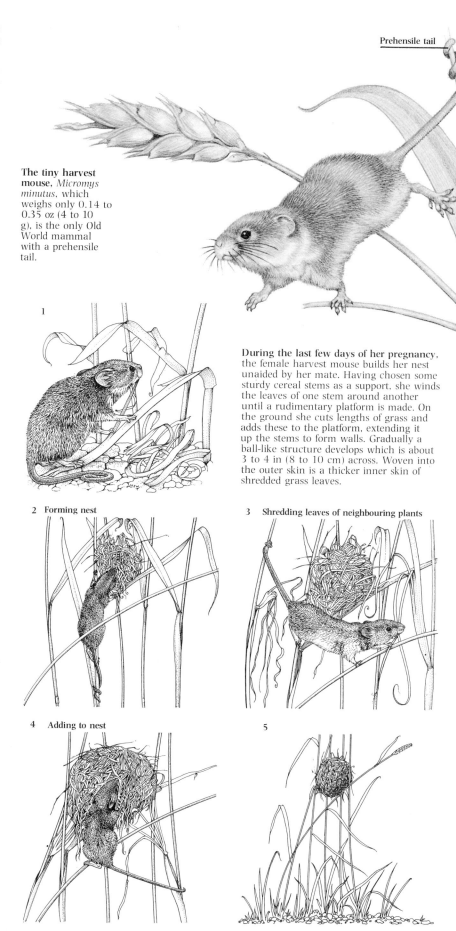

Prehensile tail

The tiny harvest mouse, *Micromys minutus*, which weighs only 0.14 to 0.35 oz (4 to 10 g), is the only Old World mammal with a prehensile tail.

1

During the last few days of her pregnancy, the female harvest mouse builds her nest unaided by her mate. Having chosen some sturdy cereal stems as a support, she winds the leaves of one stem around another until a rudimentary platform is made. On the ground she cuts lengths of grass and adds these to the platform, extending it up the stems to form walls. Gradually a ball-like structure develops which is about 3 to 4 in (8 to 10 cm) across. Woven into the outer skin is a thicker inner skin of shredded grass leaves.

2 Forming nest

3 Shredding leaves of neighbouring plants

4 Adding to nest

5

The pygmy glider, *Acrobates pygmaeus*, is the smallest tree-living marsupial and weighs 0.35 to 0.5 oz (10 to 14 g). It rarely descends to the ground, finding all that it needs high among the tree tops. The glider seeks out a tiny hole in an old tree and partially lines it with dry, chopped-up leaves. In this nest it gives birth to 3 or 4 young. Unlike other marsupials, these animals do not carry their young in a pouch but leave them in the nest while they search for nectar and small insects among eucalyptus flowers. With its combination of a small entrance and a lofty position this nest is extremely safe, and most young survive to adulthood.

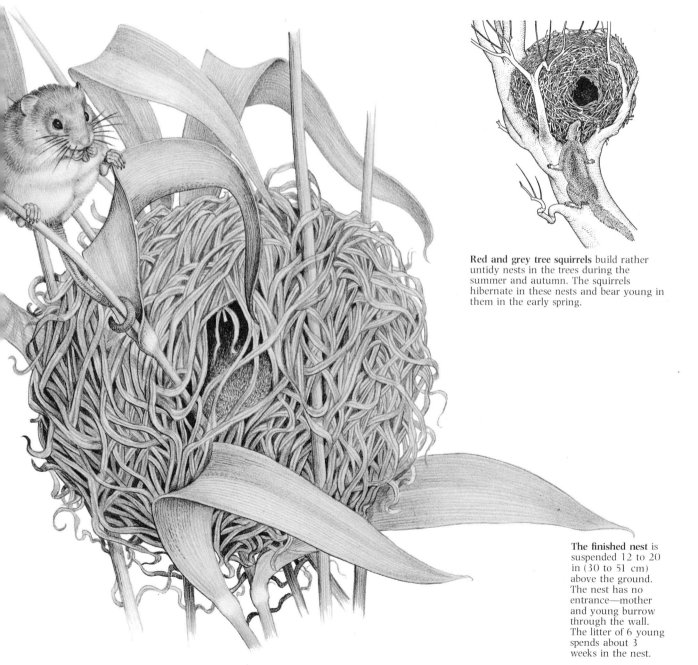

Red and grey tree squirrels build rather untidy nests in the trees during the summer and autumn. The squirrels hibernate in these nests and bear young in them in the early spring.

The finished nest is suspended 12 to 20 in (30 to 51 cm) above the ground. The nest has no entrance—mother and young burrow through the wall. The litter of 6 young spends about 3 weeks in the nest.

The moment of fertilization in sexual reproduction is rather like a marriage. Two partners, the egg and the sperm, are united and embark on a new life together. Biologically speaking, a copy of the inheritance material DNA has escaped from an ageing body, and has combined with a similar set of genes which have managed the same trick and are ready to build a sparkling new survival machine which may incorporate some innovations as a result of new gene combinations.

The body that is to be built from the single cell of the fertilized egg or ovum may be very complex. Millions of cells may have to be made and organized into tissues and organs designed to perform specialized tasks. But the genes have the know-how, gained from millions of years of change and selection. Some genes—or, more precisely, patterns of DNA organization—such as those governing fundamental life functions such as gaining energy from sugars, have existed almost since life began. Others, such as those that direct the precise building of the human foot, have a history of only a few million years. Nonetheless, genes remain an organism's direct link with the past. Ancient patterns of DNA organization are held in trust in the body to be passed on to future generations.

Unless the new animal is to be a simple, single-celled creature such as an amoeba or a bacterium, it will take time to put together. Rome was not built in a day, and nor is a man, a cow, a frog, a chicken, a beetle or an earthworm. The process of development begins with division of the fertilized egg cells. At first there may be little growth—a large cell divides to form a ball of smaller cells with no overall increase in bulk. Many animal mothers supply their fertilized eggs with a food store in the form of yolk. Like an Egyptian pharaoh left with food in his tomb for the journey into a new phase of existence, the egg's yolk reserves supply the energy and nutrients for the beginnings of a new life.

Initially, the ball of cells is solid, but as division proceeds it becomes hollow. If an ovum is provided with a great deal of yolk, this pattern of organization is distorted and the cell division may be restricted to one end (pole) of the egg, or concentrated as a disk of cells that gradually grows on the surface of a huge sphere of reserve food.

The bodies of most animals are much more complex in design than a ball or disk of cells. Nearly all creatures have an internal tube of tissue—the gut—which digests and absorbs food. Gut formation is one of the first stages in the development process. Called gastrulation, it often results from an infolding at some point on the surface of the cell ball or disk. This infolding forms a saclike organism with an outer layer of 'skin' cells and a mouth that opens into an inner layer of gut cells. In sea anemones, hydras, jellyfish and the like, development makes little further progress, but for more complex animals this is only the beginning.

For the generation of large, active organisms, a middle layer of tissue must be formed between skin and gut. Here, muscles and skeleton develop for support and locomotion, plus heart and blood vessels to provide a circulatory system. In vertebrates, active animals that require powerful muscles and high-pressure blood circulation for survival, the middle layer is particularly well developed.

Embryo development does not always follow a straightforward, obvious pattern. Free-living embryos (larvae), for example, may grow special anatomical devices to aid their own survival which are not found in the adult. They may also have a totally different lifestyle. Thus, many insects develop first as grubs or caterpillars and metamorphose dramatically into winged adults. Embryos that develop within their mother may grow elaborate feeding structures, usually a system of membranes complete with a blood circulation, to draw nourishment from the mother.

During its early development, while its essential systems are being formed, an embryo is highly vulnerable. It may be killed by sudden changes in its environment or make an easy meal for a hungry predator. Eggs that are simply shed and given no form of protection during this critical, early period are those most likely to perish.

One solution to the survival problem is to produce vast numbers of eggs. Marine animals tend to neglect their eggs but produce so many that some, at least, manage to reach adulthood. An animal which produces thousands or millions of eggs can do little for them but spawn them in a suitable place. Even the food reserves must be parsimoniously distributed among so many. Further, the laying of so many eggs may exhaust the parents, particularly the female. If all her food resources and energies are concentrated in one tremendous effort she may not live to offer any care or protection. Alternatively, the survival chances of an individual egg (and the genetic blueprint it contains) may be enhanced if fewer eggs are produced and those eggs are carefully looked after by the parents. As a rule the more negligent the parents, the more eggs produced by the female in her lifetime

In the course of evolution, natural selection has often favoured the investment of reproductive resources in just a few well-cared-for offspring. This is especially so for large, complex animals which can avoid sudden changes and create their own environmental stability despite the vagaries of the weather, day and night, heat and cold. The female of such an animal gives her developing embryos ample food, either as yolk or, if the embryo develops inside her, by direct interchange of nutrients from her blood. Parents can build nests or sheltered hiding places for their few large eggs or new-born infants. Some fish develop egg-protecting brood pouches about their bodies, or guard their eggs in their mouths or in the pouches of their gills. The female octopus hangs her eggs in a rocky recess and keeps watch over them, periodically cleaning them of debris with her tentacles and flushing clean water over them.

The ultimate in egg protection is for the female to retain eggs within her body. The eggs may be nourished by an ample yolk supply—a strategy adopted by certain insects, sharks, bony fish, lizards, snakes and many other creatures—or by a link-up with tissues of the mother. Some shark embryos start off by relying on yolk reserves, then make contact with the circulation of the mother when a placenta is formed via which nourishment and gases are exchanged. Many animal embryos, including those of most mammals, rely entirely on a placental union with the mother for their growth, as the eggs have scant food reserves. Following internal development, the young are thrust into the world as fairly advanced, smaller versions of the adult. Some, such as young hoofed mammals, and whales and dolphins, are ready to move with the group or herd almost immediately, while others, such as primates, need prolonged maternal care.

An alternative method for giving birth to advanced young is to provide an elaborate and well-protected incubator in the form of a shelled egg. Cased in a mineral-toughened shell, and generously provisioned with yolk, the embryo can grow and develop undisturbed. The shelled eggs of reptiles fulfil another role. They release the parents from the need to return to the water to lay eggs which hatch into aquatic young, and have allowed the reptiles, and the birds that followed them, the conquest of the land.

The resources invested in egg production are well insured. Reptiles often lay their eggs in carefully selected sites and cover them with sand or earth to protect them from excessive fluctuations in temperature. Birds incubate their eggs using their own bodies to warm them or protect them from the burning rays of the sun. When the eggs have hatched, some reptiles, and many birds, feed, guide and protect their offspring, easing them gently into self-sufficiency.

But when does life begin? The freely shed eggs of plaice or cod, their larval fry with pendulous yolk sacs, would certainly seem to be alive. These are young individuals with the wherewithal for survival, even though their chances are a million to one. Can the same be said of an unborn mammal foetus or half-incubated chick, both totally dependent upon the parents? Despite the moral issues these questions pose for man, biologically they have little meaning. Life is a continuous process of gene copying and rebuilding each generation's new bodies to carry the vital patterns of DNA for a while.

Every so often in the history of a set of genes they find themselves in an individual at the brink of life. Whether it is a human baby being born, a bird struggling its way out of its shell, or a butterfly emerging from its chrysalis, birth is a complex event physically, and every creature faces a crisis. Many things can go wrong and often do, but the animals in the world now have succeeded in the past. They have the genetic organization to put a new life together, at least often enough to ensure their continued presence on earth.

THE BRINK OF LIFE

The first phase of independent life is the most hazardous. Neglectful parents may lay millions of eggs in the hope that one or two will survive. Others concentrate on a small family, or deliver their offspring in protective shells.

The developing embryo

The starting point for the life of a new animal, be it a flea or a human being, is a single cell. Though often large, this cell—the fertilized egg or ovum—divides many times in the process of development. At the same time, groups of cells in the embryo become specialized into particular tissues and organs.

In vertebrate animals the early divisions of the ovum lead to the formation of a hollow ball of cells called the blastula or blastocyst. But because the eggs of most backboned creatures contain much yolk, the blastula is often distorted into a disk shape. The eggs of mammals are an exception. They have little yolk, and divide to make a spherical blastocyst. Most of the blastocyst cells do not, however, contribute directly toward the embryo's development. Instead they form a membrane which burrows into the lining of the uterus to obtain nourishment.

From the basic ball or disk of cells some cells grow inward to make a digestive tube or gut and other internal organs. In amphibians this intucking takes place through a pore, but in birds and mammals it occurs along a line of tissue known as the primitive streak. The nerve tube of vertebrate embryos, which will become the spinal cord, develops quite early on and becomes thickened at the front end to form the brain. Also among the first organs to arise are the heart and circulatory system. These are vital for supplying growing tissues with food and oxygen. As the embryo begins to take on a recognizable body form, body muscles appear, plus buds that will become limbs.

While it gradually takes shape, the embryo of a vertebrate partly lives out its ancestral history. At an early stage a mammal embryo is decidedly fishlike and even has partially formed gill slits that later disappear or are changed into other structures. This apparent process of recapitulation is only part of the story, for embryos develop structures never seen in the adult of any ancestor but which suit their own particular needs.

Perhaps the most fascinating question of embryo development is, why should one cell in a blastula be destined to become part of the nervous system and another, which looks identical, part of the gut? It seems that the position of any cell in the developing embryo has much to do with its fate. The destiny of each cell is already broadly mapped out at the blastula stage, so that each cell's placement determines the migrations it will make and how it will become specialized. Successful embryo development, like poetry, demands 'the right words in the right order'.

PLATYFISH

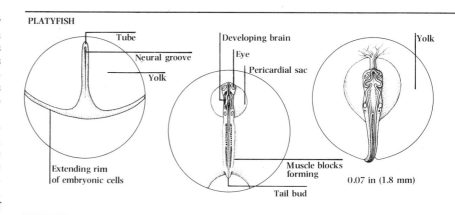

FROG, *Rana pipiens*

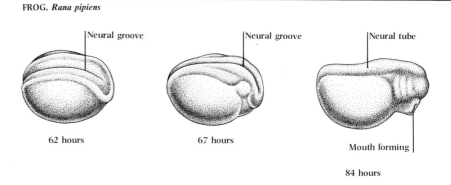

CHICK

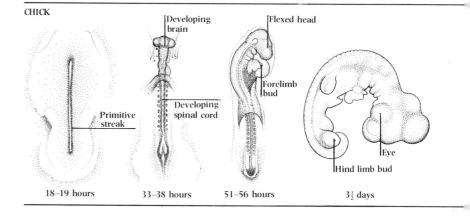

MAN

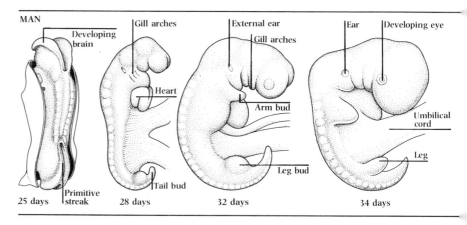

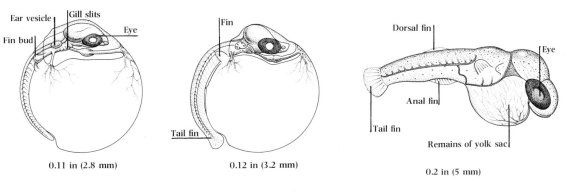

Ear vesicle | Gill slits
Fin bud
Eye

0.11 in (2.8 mm)

Fin

Tail fin

0.12 in (3.2 mm)

Dorsal fin
Eye

Anal fin

Tail fin

Remains of yolk sac

0.2 in (5 mm)

The development of the fish's central nervous system begins with an axis of cells which is partly a groove, partly a tube. The brain and eyes begin to develop. The tail then becomes mobile and the embryonic heart forms. By the time the embryo is almost ready to hatch all the fin buds are evident and the yolk sac is much reduced.

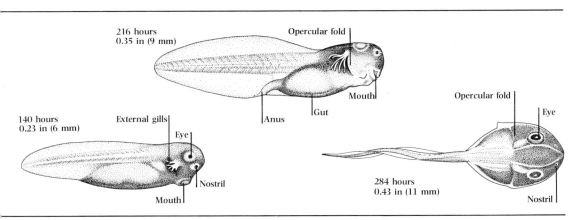

216 hours
0.35 in (9 mm)

Opercular fold

Mouth

Anus | Gut

140 hours
0.23 in (6 mm)

External gills

Eye

Nostril

Mouth

Opercular fold

Eye

284 hours
0.43 in (11 mm)

Nostril

At 6 hours the neural groove is apparent and dilating in the region where the brain is to form. Around 70 hours the neural tube seals so that all the central nervous system is enclosed. At 84 hours the tail bud starts to form. By about 140 hours the external gills become obvious, and the embryo hatches into a tadpole at this stage.

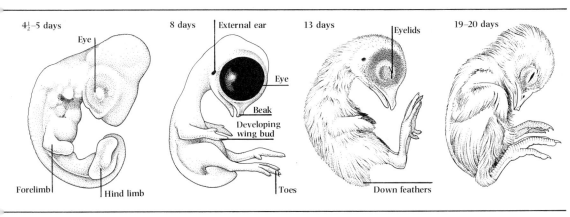

$4\frac{1}{2}$–5 days

Eye

Forelimb | Hind limb

8 days

External ear

Eye

Beak

Developing wing bud

Toes

13 days

Eyelids

Down feathers

19–20 days

After 16 hours of incubation, movement of cells through the primitive streak area is under way. In 34 hours the nerve cord and brain are well-developed. At 48 hours the chick turns on its side and the circulation supplies nourishment from the yolk. At 9 days the embryo has a large head and eyes and developing limbs. It hatches at 21 days.

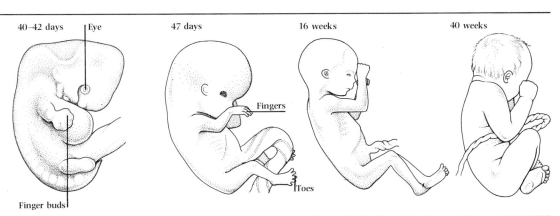

40–42 days

Eye

Finger buds

47 days

Fingers

Toes

16 weeks

40 weeks

A prime concern is the development of the embryo's heart and blood circulation; the major blood vessels are complete by the age of 4 weeks when the embryo is only 0.18 in (4.5 mm) long. By 8 weeks the foetus has a well-formed head and limbs, and bone formation has begun. Growth is now rapid; by 10 weeks the foetus is nearly 2 in (50 mm) long.

The making of an egg

Within the enclosed and protected environment of an egg the early stages of an animal's development can be completed, so that the young creature hatches out as a well-formed but smaller version of its parents. In the evolution of land-living vertebrates the shelled egg was a major breakthrough, for they no longer required an aquatic environment for larval life.

A bird's egg is a complex structure consisting not only of the female sex cell or ovum but also of a number of membranes, nutritive layers and the shell. These extra layers are added to the ovum after its release from the ovary and during its journey down the oviduct.

Compared with other cells the bird's ovum is huge because it contains the nutrient-rich yolk. This yolk is not structureless, but consists of a central white core around which are alternate layers of yellow and white yolk. (The white layers are much narrower than the yellow and contain less fat and pigment.) The central white core extends as a thin rod to the edge of the egg. Here it makes contact with the germinal spot—

the place where, in a fertilized egg, embryo development begins.

Surrounding the egg yolk is a vitelline membrane made up of two layers. The inner layer is produced in the ovary, while the outer, more finely fibrous layer is formed when the egg enters the top part of the oviduct. The egg proper is suspended in a pool of egg white, which is about 85 per cent water but also contains proteins, called egg albumen.

Anchoring the ovum-containing yolk in its watery surrounds are fibrous strands of albumen, the chalazae. Attached at the opposite ends of the yolk and parallel to the egg's long axis, the chalazae are twisted—one set clockwise, the other anti-clockwise—which shows that the egg rotates during albumen formation.

Enclosing the egg white are two parchmentlike shell membranes made mostly of keratin, the main structural protein of a bird's feathers. The inner and outer shell membranes are in close contact except at one point—usually at the broad end of the egg—where they are separated to form an air space.

Finally, a shell is secreted round the egg. Crystals of calcium carbonate are the shell's main components, but it also comprises magnesium, sodium and potassium-containing compounds. Mature female birds have reservoirs of calcium for shell making in their leg bones. The shell is sheathed in a protein cuticle and perforated by thousands of tiny pores which allow the embryo to breathe.

Usually, shell thickness varies with egg size. The tiny eggs of the hummingbirds have the thinnest shells, those of the extinct elephant birds, *Aepyornis*, the thickest. Many eggs are coloured, often for camouflage. Desert-nesting coursers and sandgrouse have pale eggs, while the Temminck's courser, *Cursorius temminckii*, which nests on burnt ground, produces eggs that are nearly black. Guillemot eggs show a remarkable range of speckled patterning, even in one colony. These birds nest on flat rocks, and eggs from different females may roll around and get muddled up, but each bird can recognize its own eggs by their distinctive markings.

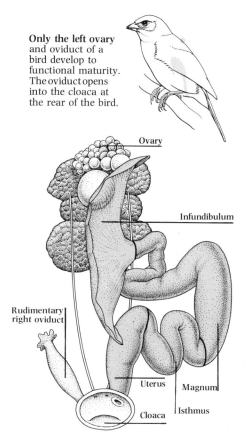

Only the left ovary and oviduct of a bird develop to functional maturity. The oviduct opens into the cloaca at the rear of the bird.

Ovary

Infundibulum

Rudimentary right oviduct

Uterus

Magnum

Cloaca

Isthmus

Different regions of the oviduct contribute secretions to the egg: the infundibulum, a funnel which engulfs the released ovum; the magnum and isthmus with secretory glands; and the uterus or shell gland.

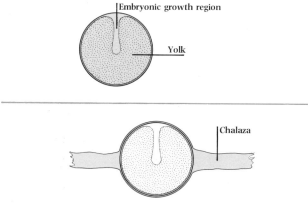

Embryonic growth region

Yolk

Chalaza

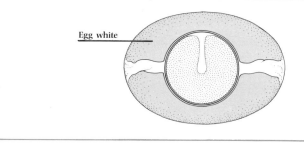

Egg white

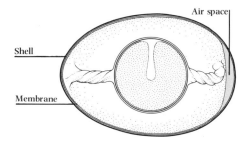

Air space

Shell

Membrane

The ovum enters the infundibulum complete with the food reserves of the yolk. As it passes down the magnum, albumen (egg white) is added in layers. The chalazae constitute one layer. These are dense fibres of albumen firmly attached to the surface of the yolk at its equator. Their outer layers interlace with fibres in the rest of the albumen to stabilize the yolk's position.

The inner and outer shell membranes are added as the egg passes through the isthmus. The outer membrane is thicker and has 3 layers. In the uterus, water enters the egg through the membranes. This increases the amount of thin or liquid white, and effectively doubles its weight. In the uterus the egg is rotated, coiling the chalazae. The shell itself is secreted in the uterus and any colouring is added in the last few hours before the egg is laid.

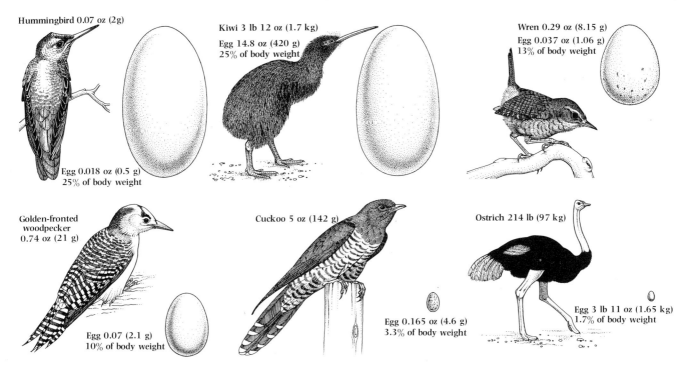

Hummingbird 0.07 oz (2g)
Egg 0.018 oz (0.5 g)
25% of body weight

Kiwi 3 lb 12 oz (1.7 kg)
Egg 14.8 oz (420 g)
25% of body weight

Wren 0.29 oz (8.15 g)
Egg 0.037 oz (1.06 g)
13% of body weight

Golden-fronted
woodpecker
0.74 oz (21 g)
Egg 0.07 (2.1 g)
10% of body weight

Cuckoo 5 oz (142 g)
Egg 0.165 oz (4.6 g)
3.3% of body weight

Ostrich 214 lb (97 kg)
Egg 3 lb 11 oz (1.65 kg)
1.7% of body weight

For many birds the weight of their egg is 7 to 10 per cent of the adult body weight, although large birds produce relatively smaller eggs. An extreme example is the ostrich, which lays the largest egg of any living bird—it weighs 3 lb 11 oz (1.65 kg) —but that egg is only 1.7 per cent of its body weight. The tiny hummingbird lays an egg weighing only 0.018 oz (0.5 g), which is 25 per cent of the weight of the parent bird. The kiwi, although a large bird, lays an egg weighing 25 per cent of its body weight. The cuckoo produces an egg only 3.3 per cent of its own body weight.

The chalazae are attached to opposite poles of the yolk and have a cloudy spiral appearance. Their function is to stabilize the egg cell in its position in the surrounding egg white.

The germinal spot contains the fused nuclei of a fertilized egg, which develop into a embryo.

The outer shell membrane is closely attached to the inner surface of the shell.

The shell, made of calcium carbonate, is porous to allow the chick to breathe.

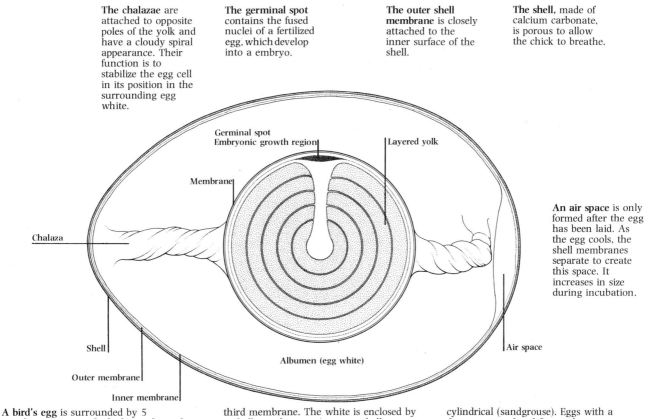

Germinal spot
Embryonic growth region
Layered yolk
Membrane
Chalaza
Shell
Outer membrane
Inner membrane
Albumen (egg white)
Air space

An air space is only formed after the egg has been laid. As the egg cools, the shell membranes separate to create this space. It increases in size during incubation.

A bird's egg is surrounded by 5 membranes, some of which are themselves composed of a number of layers. The true egg cell with its yolk is surrounded by a double vitelline membrane. This yolk is suspended in the egg white which is the third membrane. The white is enclosed by 2 shell membranes. A porous shell, composed mainly of calcium carbonate, protects the whole structure. The shape of eggs varies from almost spherical (some owls and kingfishers) to elongate and cylindrical (sandgrouse). Eggs with a distinct pointed end fit snugly into a nest with the point toward the centre. This arrangement occupies the minimum space and allows the brooding parent to cover the eggs easily.

Laying and incubating eggs

Birds use the heat generated by their warm-blooded bodies to provide a stable temperature for the early development of their young. To pass on this heat they incubate their eggs, and during incubation the average temperature of each egg is about 34 degrees centigrade, some 6 or 7 degrees below the bird's body temperature. At incubation time, bird feathers are a doubtful asset, for they provide the body with excellent insulation, and so some must be shed if the bird is to use its body heat to warm the eggs. This is why many birds have naked brood patches on their breasts.

In a nest the temperature at the top—nearest the bird's body—is higher than at the bottom, so the parent regularly turns the eggs to ensure even warming. Where nests are rudimentary or non-existent, other methods of incubation are employed. Boobies and gannets warm their eggs by standing on them, the blood circulation in their webbed feet providing the essential heat. Penguins do the opposite, balancing their eggs on top of their feet. Emperor penguins also have a fold of belly skin which snugly envelops and warms the egg.

Because most birds produce a clutch of eggs—laid at about 24 hour intervals—the parents have a problem. Should they start incubation as soon as the first egg is laid, in which case the young will hatch at different times, or suspend incubation until the clutch is complete, thus roughly synchronizing hatching but exposing unprotected eggs to predators and the weather? Birds that follow the first policy may have some flexibility of family planning. In times of food shortage, late eggs are neglected, but in plentiful seasons parents can try to rear the whole brood. The second scheme has the advantage that parents can switch smoothly from incubating behaviour to caring for their newly hatched young.

Whatever the method, incubation duties are usually shared between the two parents, who may observe a strict schedule. In species that nest in the open the female generally specializes in incubation while the male forages for food. This reduces the amount of activity around the nest and helps to keep it hidden from predators. In a few birds, including kiwis, emus and emperor penguins, only the male incubates eggs.

The work of breaking out of the egg is usually left to the chick. Young birds develop two tools to aid their bid for freedom—an egg tooth on top of the upper beak for piercing the shell, and a hatching muscle along the upper side of the head and neck to help the tooth to do its work. Both these structures disappear soon after hatching.

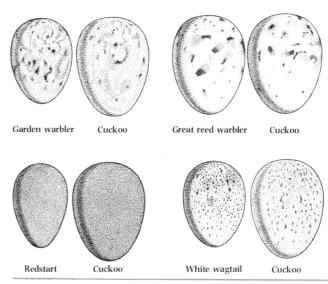

Garden warbler   Cuckoo   Great reed warbler   Cuckoo

Redstart   Cuckoo   White wagtail   Cuckoo

If a cuckoo's egg looks too different in size and colour to the host's egg it may be hurled from the nest. In many cases the eggs of the European cuckoo have come to resemble those of the host commonly used. In Finland, the cuckoo's chief hosts are the redstart and the whinchat. Both these birds lay blue eggs, so the cuckoo also lays blue eggs. In southern England, however, the cuckoo victimizes many hosts with dissimilar eggs, so its egg is generalized in colour.

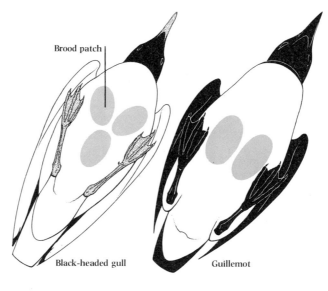

Brood patch

Black-headed gull   Guillemot

Close contact between the parent's bloodstream and the egg's shell is needed for incubation. Many birds develop bare areas of skin just before incubation to increase the transfer of heat. The skin appears inflamed because it is richly supplied with blood vessels. Grebes and pigeons have a single patch; auks, skuas and many shore birds have 2 lateral patches, and gulls and waders 3. Feathers are lost as a result of hormonal changes, but ducks and geese pluck out their own feathers.

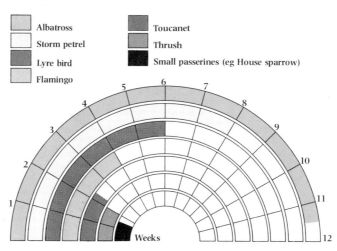

Albatross
Storm petrel
Lyre bird
Flamingo
Toucanet
Thrush
Small passerines (eg House sparrow)

Weeks

The eggs of large birds generally need a longer incubation than those of small birds. Small passerines (perching and singing birds) incubate for 10 to 14 days; hens, 21 days; owls, 26 to 36 days; petrels, 38 to 56 days and albatrosses, 63 to 81 days. There are exceptions. The tiny hummingbird incubates for 16 days, a period undercut by many larger birds, and the ostrich for only 42 days.

The European cuckoo, *Cuculus canorus*, is not the only bird to use foster parents. Members of 5 different families of birds are brood parasites.

The cuckoo usually uses smaller birds as foster parents. If it is impossible for the female cuckoo to sit on the nest to lay, she can drop her egg into the nest while hovering above it. The egg has a tough shell to survive the fall. The cuckoo lays in a few seconds—most birds take a few minutes or even hours.

Brood parasites are the tricksters of the bird world. They put their eggs in other birds' nests so that the work of incubation and feeding is done for them. The female cuckoo usually lays her egg in a freshly completed clutch to ensure maximum incubation. It hatches in $12\frac{1}{2}$ days, usually earlier than the host's eggs. About 10 hours after a cuckoo is hatched it may push other eggs or young out of the nest so that it may have all the available food.

Strategies for birth

There are two main strategies for bearing young: producing large numbers of eggs, many of which will be wasted because the parents cannot protect and nurture them, or, reducing the numbers of eggs and improving their chance of survival by caring for them. The most refined form of parental care exists among those species that keep the developing eggs in their bodies and give birth to fully formed young. This method of giving birth to young is not the sole province of placental mammals—it has evolved in some representatives of nearly every major animal group.

Reproduction in which all or most of the development of the embryo takes place after the egg has been laid is known as oviparity. All birds and many fish, reptiles and invertebrates use this system. Embryonic development inside the body of the non-mammalian female is termed ovoviviparous or viviparous. In the former, eggs are produced with sufficient food reserves (yolk) for growth and simply retained inside the body, usually without any additional nutrients provided by the mother. Often the eggs develop within the oviduct of the female, but any cavity of the body may be used. Some fish use their mouths, others the gill cavity. The seahorse male develops a special brood pouch in which he looks after the eggs laid by the female. The Surinam toad, *Pipa pipa*, of equatorial South America, makes use of an unusual site. After the female has laid her eggs and they have been fertilized, the male assists in placing them on her back. Her skin then swells to enclose the eggs. She keeps the young in these pockets on her back until they pass through the tadpole stage to emerge as small toads. In these cases internal fertilization of the eggs is not necessary.

Females of viviparous species produce eggs with little or no yolk, which then draw nourishment from the mother as they grow. Nutrients may simply be secreted into the chamber in which the embryos are developing, or the embryo may be attached to a special 'milk gland' within the oviduct. The most specialized system is one in which close contact is established between the embryonic and maternal circulations so that food and excretory and respiratory exchanges can take place directly. The formation of a structure known as a placenta for this purpose is not restricted to mammals; it is also found in other vertebrates, notably certain sharks and snakes.

Those fish that give birth to live young instead of simply shedding eggs and sperm into the water, must have some means of internal fertilization. Mature male sharks have copulatory organs, claspers, which are modifications of the pelvic fins. Each clasper has a cartilaginous skeleton, and a groove runs its length from a gland at the base. To copulate, the two grooves are brought together, the claspers are placed into the female's cloaca, and sperm is transferred.

The distinction between ovoviviparous and viviparous development is somewhat artificial; fish, snakes, scorpions and insects show examples of both. In fact, in some species embryonic development involves both an ovoviviparous and viviparous stage. In the shark, *Mustelus canis*, the placenta develops to nourish the young only after several months of ovoviviparous existence.

Some form of internal embryonic development is found in every class of vertebrates with the exception of birds. Presumably bird flight would be hindered by the additional weight of a developing embryo.

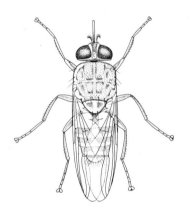

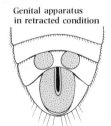

The tsetse fly, *Glossina*, is an example of a viviparous insect. One larva develops at a time in the oviduct. Male tsetses have genitals on the underside of the abdomen.

Genital apparatus in retracted condition

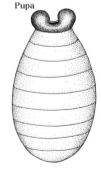

Larva   Pupa

EMBRYOS WITHIN EMBRYOS
The parasitic fluke, *Gyrodactylus*, lives on the skin of freshwater fish. Eggs pass from the fluke's ovary to the brood chamber. The eggs lack yolk, but only one develops and gains nourishment from sister eggs. Quite early in development a second embryo appears inside the first, and a third inside the second, and so on. The first embryo emerges still containing the others.

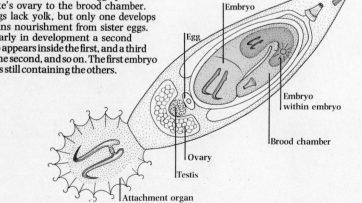

Embryo

Egg

Embryo within embryo

Brood chamber

Ovary

Testis

Attachment organ

The tsetse larva is nourished inside the mother by a milk gland which opens into a nipple close to its mouth. It is about 0.4 in (10 mm) long when born. It passes through a pupal stage of 17 days to 3 months before the adult emerges.

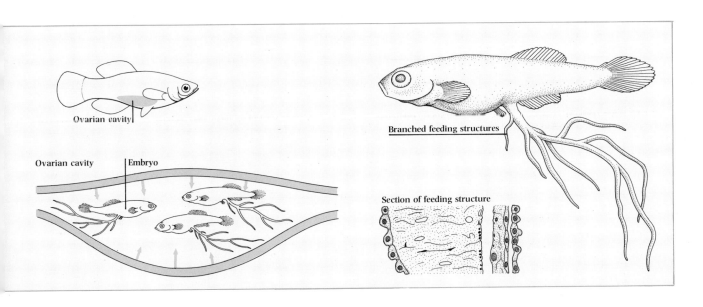

Ovarian cavity

Ovarian cavity   Embryo

Branched feeding structures

Section of feeding structure

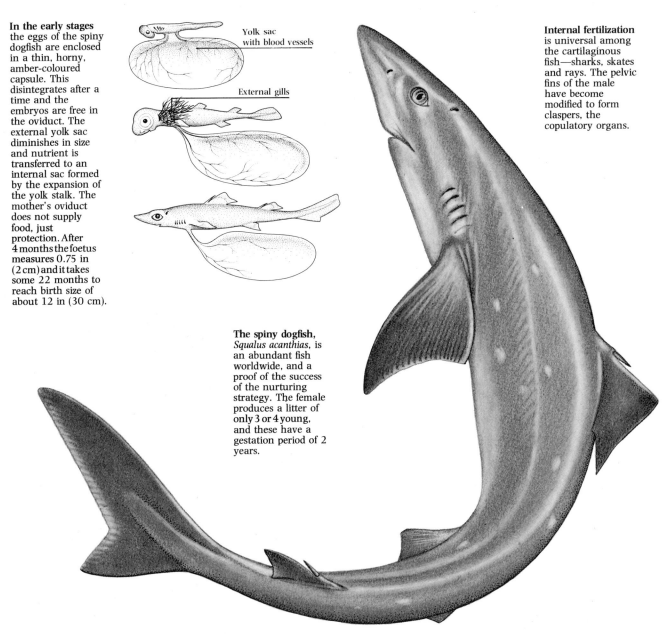

In the early stages the eggs of the spiny dogfish are enclosed in a thin, horny, amber-coloured capsule. This disintegrates after a time and the embryos are free in the oviduct. The external yolk sac diminishes in size and nutrient is transferred to an internal sac formed by the expansion of the yolk stalk. The mother's oviduct does not supply food, just protection. After 4 months the foetus measures 0.75 in (2 cm) and it takes some 22 months to reach birth size of about 12 in (30 cm).

Yolk sac with blood vessels

External gills

Internal fertilization is universal among the cartilaginous fish—sharks, skates and rays. The pelvic fins of the male have become modified to form claspers, the copulatory organs.

The spiny dogfish, *Squalus acanthias*, is an abundant fish worldwide, and a proof of the success of the nurturing strategy. The female produces a litter of only 3 or 4 young, and these have a gestation period of 2 years.

Larvae—the halfway house

Embryonic young which must survive independently while maturing, larvae, are produced by many animals. The appearance, habitat and food sources of larvae are often distinctly different from those of their parents. In addition, their locomotory and feeding apparatus, as well as their behaviour patterns, may be dissimilar. As a result, a single species may exploit two or more sorts of environment, and the immature young are not in direct competition with the adults for resources. In these cases the larva usually undergoes a dramatic metamorphosis to the adult form.

In general, insect larvae are less mobile than the adults. Wormlike larval caterpillars slowly feed on plant leaves and stems with their cutting mouthparts, but the adult butterflies are highly mobile and have an elongated tubular 'tongue' for sucking nectar and the juices of overripe fruit. The larvae primarily feed and grow, while the adults concentrate on mating.

The larvae of stoneflies, dragonflies, and mayflies live in freshwater streams as nymphs. Breathing through gills and feeding on small aquatic organisms, they may live for several years, surviving at least one winter before emerging as fully formed winged adults. In fact, the larval stage dominates the life cycle of these animals. The life span of the adult form is brief—only a few hours in some mayflies, just long enough to mate and produce eggs.

Among marine molluscs and crustacea, it is the larvae that are mobile and responsible for dispersal of the species. Adult limpets have restricted mobility, and barnacles and mussels anchor themselves firmly to rocks, but they produce free-swimming larvae which, carried by tides and currents, can settle in new and perhaps richer colonies.

Like mayflies, amphibia have aquatic larvae. Some amphibia fail to metamorphose into adults, and the larvae become sexually mature, a phenomenon known as neoteny. Mud puppies of the genus *Necturus*, for example, live entirely as aquatic larvae.

The tunicates, or sea squirts, may seem unlikely candidates for a position close to the vertebrate ancestral stock. They are small, inactive, marine, filter-feeding animals, enclosed within leathery tunics. Some live attached to a rock or plant, others float freely in the sea. But they have active, mobile larvae with a tough incompressible rod, the notochord, along the length of their muscular tails, and a hollow dorsal nerve cord. Both of these are vertebrate features. Neoteny among such larvae may well have played a crucial role in the evolution of vertebrate animals.

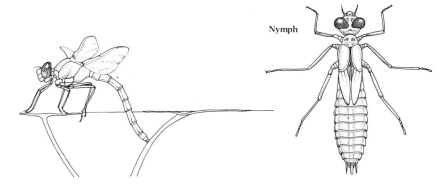

A female dragonfly lays her eggs in or near water. Many species have rounded eggs which are dropped into the water or attached to aquatic plants.

The first part of a dragonfly's life is spent as an aquatic nymph. It has gills in its rectum and to 'breathe' it pumps water in and out of the rectum through the anus.

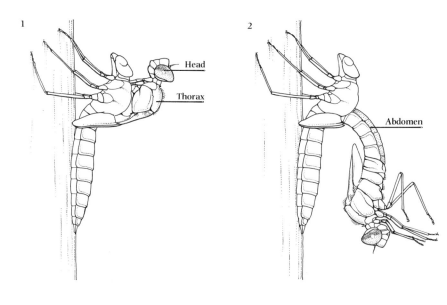

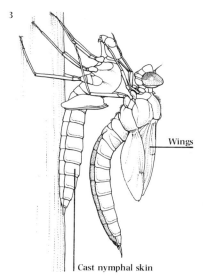

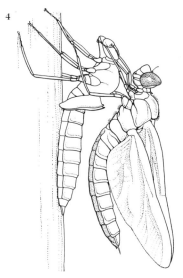

A fully grown nymph crawls out of the water on to a plant or rock to undergo its metamorphosis to the adult form. It emerges from its nymphal skin, which splits along the back. First it withdraws its head and thorax, then the legs and wings.

It waits, head hanging down, while its legs attain full strength and range of movement. Finally it withdraws the abdomen. The abdomen and wings take about half an hour to become fully extended.

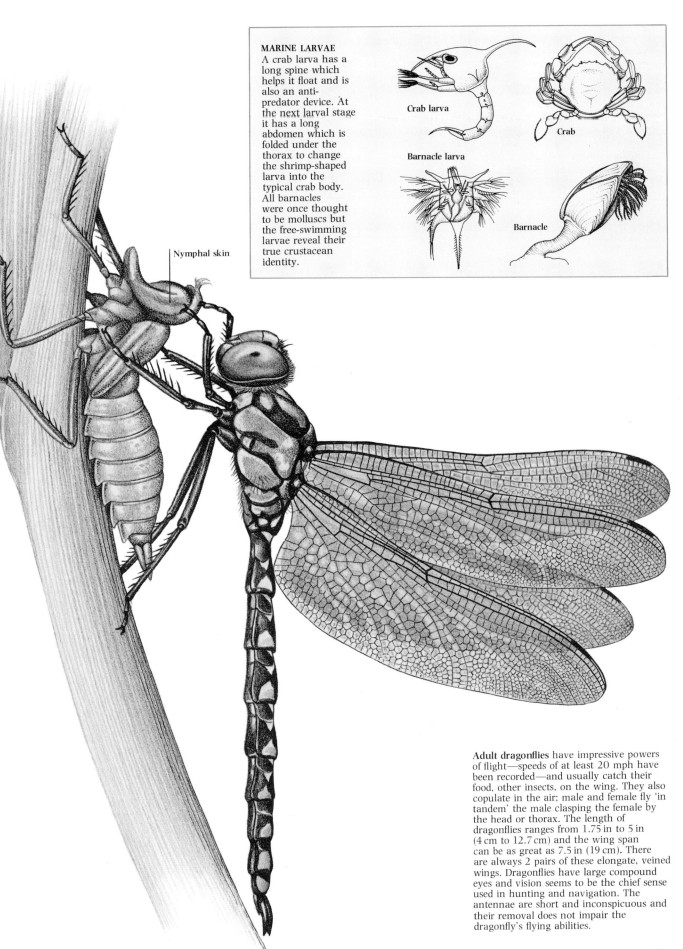

MARINE LARVAE
A crab larva has a long spine which helps it float and is also an anti-predator device. At the next larval stage it has a long abdomen which is folded under the thorax to change the shrimp-shaped larva into the typical crab body. All barnacles were once thought to be molluscs but the free-swimming larvae reveal their true crustacean identity.

Crab larva

Crab

Barnacle larva

Barnacle

Nymphal skin

Adult dragonflies have impressive powers of flight—speeds of at least 20 mph have been recorded—and usually catch their food, other insects, on the wing. They also copulate in the air; male and female fly 'in tandem' the male clasping the female by the head or thorax. The length of dragonflies ranges from 1.75 in to 5 in (4 cm to 12.7 cm) and the wing span can be as great as 7.5 in (19 cm). There are always 2 pairs of these elongate, veined wings. Dragonflies have large compound eyes and vision seems to be the chief sense used in hunting and navigation. The antennae are short and inconspicuous and their removal does not impair the dragonfly's flying abilities.

Animals at birth

Warm-blooded animals—birds and mammals—nourish and protect their embryos either outside the body in an egg, or inside in the uterus. They give birth to young that are smaller versions of themselves, although there is a great variation in the degree of their development. Some animals are altricial, that is they are born blind and helpless, without fur or feathers, to species that provide a well-protected nest, burrow or other shelter. Others are precocial: they are born in the open, well-formed, with acute senses and capable of fending for themselves almost immediately.

The passerines (singing and perching birds) build elaborate and well-guarded nests for their offspring, which hatch with their eyes closed, almost or wholly naked and unable to leave the nest. They are completely dependent upon their parents for food. Newly born hawks, falcons and vultures are also unable to leave the nest and must be fed by their parents, but they are covered with down and emerge with their eyes open. Birds such as pheasants, partridges, fowl, ducks and geese, on the other hand, hatch as precocial young and can immediately follow their parents to find food. The young of a clutch must all hatch at the same time, as any late hatchling will be left behind. The offspring of the brush turkey represent extreme examples of precocial young. The mother incubates her eggs in a mound of rotting vegetation. The young hatch strong enough to dig their way to the surface of the mound and are able to fly within 24 hours.

Mammals which give birth in burrows or other secure sites, insectivores, rodents, carnivores and rabbits, generally produce altricial young that are relatively small. The cub of the brown bear, *Ursus arctos*, for example, weighs only about 18 ounces (500 grams). This may be due in part to the fact that bear cubs are born in winter when the mother is in a den, where it would be difficult for her to cope with highly active young. But even bears that live in the tropics and do not have a 'denning up' period produce tiny offspring, indicating their northern origins.

Ungulates (hoofed mammals, such as horses, deer and antelope), whales, dolphins and hares give birth in the open and their offspring are large and precocial. A new-born ungulate, for example, is a tenth the weight of its parent, and a dolphin is 10 to 15 per cent. Only a few are born in a litter, often just one, and the gestation period is long: 11 months for a horse and blue whale, and 14 to 15 months for a giraffe.

NEW-BORN YOUNG
Hares live in the open and their young are born with their eyes open and their bodies covered with fur. In contrast, rabbits build complex burrows and their young are well sheltered. Baby rabbits are much smaller than young hares, and are born naked, blind and helpless.

Hare

Lapwing

Rabbit

A newly hatched lapwing is open-eyed, downy and able to walk. A blackbird is naked, blind and highly dependent on its parents.

Blackbird

Some 20 minutes after birth, a giraffe can stand, albeit uncertainly, and within an hour begins to suckle. Giraffes, *Giraffa camelopardalis*, are typical ungulates in that they give birth to such highly developed young. Calves are vulnerable to predators and must quickly learn to move with the herd as they have no other protection. Usually a mother has a single calf, although twins have been known. A new-born calf weighs 88 to 154 lb (40 to 70 kg) and stands about 6.5 ft (2m) high. Some mothers suckle young for 9 months or more but giraffes start to feed on vegetation when 2 or 3 weeks old.

Despite its appearance, the giant panda, *Ailuropoda melanoleuca,* may not be closely related to bears but to the family Procyonidae which includes the American raccoon. Like bears, however, the panda gives birth to extremely small young. At birth a panda is blind, toothless and helpless; it weighs only about 5 oz (141 g). At 250 lb (113 kg) a female panda is about 800 times the weight of her baby. In the early weeks the mother cradles her baby; it does not crawl for 3 or 4 months.

Young pandas grow rapidly, and although they are so tiny when born, by the age of 8 weeks they are more than 20 times their birth weight. Growth figures are based on baby pandas observed at Peking Zoo.

The primitive mammals

Although monotremes, which lay eggs, and marsupials, which bear live undeveloped young, are the most primitive living mammals, it is in the care that they give their young that they are most like their more advanced relatives. Monotremes do not have well-defined mammary glands, so the young have to lap up the milk that oozes out over the mother's abdomen. The terrestrial echidna carries its egg with it in an incubation groove on the abdomen. Once hatched, the youngster enjoys its mother's warmth until its growing spines irritate her. The aquatic platypus digs a burrow for its eggs, incubates them and then feeds the young.

The majority of marsupials have an abdominal pouch into which the mammary glands open. Small, mouselike marsupials, however, have only a pair of fleshy flaps enclosing the mammary gland area. In most species the opening of the pouch faces forward, but in bandicoots, koalas and wombats it faces backward. Except for antechinuses, which have a nipple for each youngster, female marsupials usually have more nipples than their young need.

The brown bandicoot, *Isoodon obesulus*, for example, has eight nipples but never bears more than four young. As the litter grows, the nipples that are used enlarge until at the time of weaning they become too big for a new batch of young to suckle. Consequently, the next litter of bandicoots occupies the four smaller nipples. As these expand, the other previously large nipples gradually shrink until they can be used by a third litter. So, litter can follow litter without a break. Since antechinuses produce only one litter a year, it does not matter that all their nipples enlarge simultaneously.

Kangaroos may suckle two joeys at a time; the younger is confined to the pouch while the older stays outside, returning to the pouch only for feeding. The mammary gland that the older joey uses produces richer milk than the one from which the younger feeds. But as the young joey grows, the chemical composition of the milk gradually changes.

Marsupials are often said to be inferior to the placentals because the latter have replaced them in many parts of the world. In fact, even in Australia, the delicate balance has been upset by man, who has introduced placental carnivores such as cats and dogs. As a consequence, the predation pressure has proved too great for many species to withstand. The previously low level of marsupial predation may partially account for the low reproductive rates of Australian marsupials.

The reptilian ancestry of the echidna, *Tachyglossus aculeatus*, is evident in its reproduction. Although a mammal, it has retained the reptilian ability to lay a soft-shelled egg. Somehow this egg is moved from the cloaca to an incubation groove on the echidna's abdomen. The soft shell is covered with sticky mucus which holds the egg on to the abdomen.

Incubation groove

A kangaroo's pregnancy lasts only about 33 days. During this time a fertilized egg is developing in the reproductive tract. A couple of hours before birth the mother pays great attention to the cleanliness of her pouch. She bends her head down and nuzzles at the pouch lining, then as birth becomes imminent she squats down with tail bent forward between her hind legs. Holding the pouch with her forepaws, she salivates profusely and licks the pouch inside and out. She may still be licking at the moment of birth.

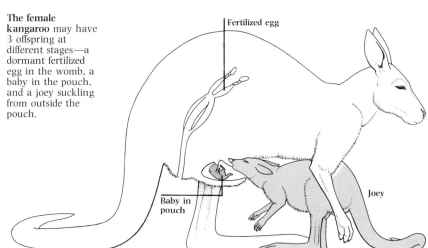

The female kangaroo may have 3 offspring at different stages—a dormant fertilized egg in the womb, a baby in the pouch, and a joey suckling from outside the pouch.

Fertilized egg

Baby in pouch

Joey

The reproductive system of the kangaroo is well suited to its arid habitat. Having mated, a number of eggs are fertilized and all but one lie dormant in the uterus. After 33 days the developing embryo is born. If there is sufficient good food available the mother will suckle the baby in the pouch for 4 months and for up to a year outside the pouch. If the rains do not come or food is otherwise inadequate, the female ejects the joey in her pouch and allows a dormant embryo to develop.

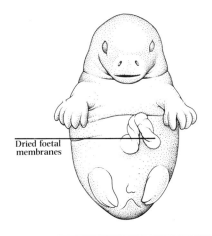

Dried foetal membranes

When it hatches, the baby echidna is 0.5 in (1.25 cm) long. The mother's milk flows from a mass of mammary glands but there is no special nipple; the baby feeds by drawing up a projection of skin into its mouth. This projection, or papilla, has grooves from which the baby sucks the milk. Once the young echidna's spines develop, its mother no longer carries it.

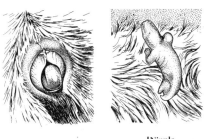

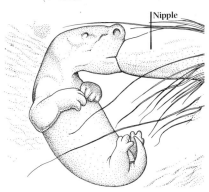

Nipple

The new-born kangaroo or joey is 0.75 in (2 cm) long and is 1/30,000 of its mother's weight. It moves unaided from the cloaca up to the pouch using the well-developed front claws. The journey takes about 3 minutes.

Inside the pouch the joey draws a nipple into its mouth. The nipple responds by swelling so that it cannot readily be dislodged. As the baby grows the nipple thickens and elongates to about 4 in (10 cm) at the time of weaning. It takes some months to shrink to its original size so as to be available to a new joey.

Brush-tailed possum

Epipubic bone

Pelvic girdle

A marsupial's pouch is a tough elastic structure that must support considerable weight—when a joey leaves the pouch it may weigh 20 lb (9 kg). To support the rim of the pouch, a pair of pelvic girdle bones, the epipubic bones, develop.

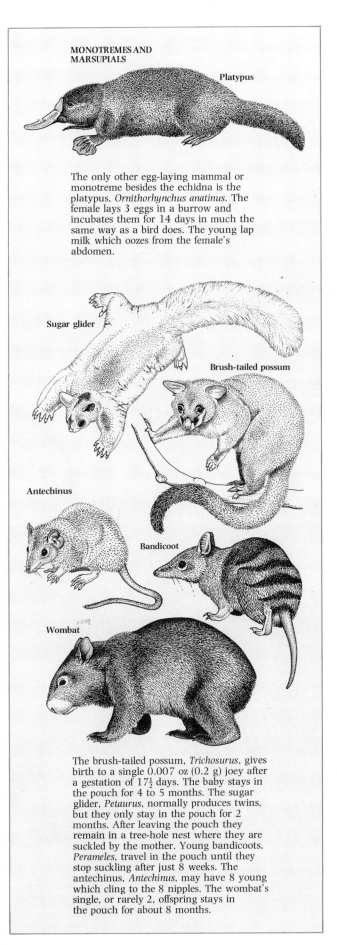

MONOTREMES AND MARSUPIALS

Platypus

The only other egg-laying mammal or monotreme besides the echidna is the platypus, *Ornithorhynchus anatinus*. The female lays 3 eggs in a burrow and incubates them for 14 days in much the same way as a bird does. The young lap milk which oozes from the female's abdomen.

Sugar glider

Brush-tailed possum

Antechinus

Bandicoot

Wombat

The brush-tailed possum, *Trichosurus*, gives birth to a single 0.007 oz (0.2 g) joey after a gestation of $17\frac{1}{2}$ days. The baby stays in the pouch for 4 to 5 months. The sugar glider, *Petaurus*, normally produces twins, but they only stay in the pouch for 2 months. After leaving the pouch they remain in a tree-hole nest where they are suckled by the mother. Young bandicoots, *Perameles*, travel in the pouch until they stop suckling after just 8 weeks. The antechinus, *Antechinus*, may have 8 young which cling to the 8 nipples. The wombat's single, or rarely 2, offspring stays in the pouch for about 8 months.

A life-support system

By any criteria, the placental mammals from mice to men are highly successful groups of animals. One of the keys to their ascendancy is undoubtedly their ability to nurture and protect their offspring for long periods during early development inside the mother.

In humans, the ovum is fertilized in the Fallopian tube joining ovary and uterus. By the time it enters the uterus some four days later, it has become a hollow ball of cells called a blastocyst. Only one region of the ball, the embryoblast, is destined to form the embryo.

Next, the blastocyst burrows into the lining of the uterus, destroying uterine cells as it goes and obtaining nourishment from them. The trophoblast region of the blastocyst does the burrowing and then develops fingerlike projections, villi, which increase the area available for absorbing nutrients. The blastocyst becomes completely embedded in the uterus lining. As the trophoblast continues its invasion, it inevitably comes into contact with blood vessels of the mother. By this time the trophoblast, now called the chorion, has also developed a blood circulation. This becomes linked to the circulation of the developing embryo which begins to derive life support via its blood system about four weeks after fertilization.

The villi of the chorion continue to grow and expand, making further intrusions into the maternal blood vessels. To complete the placenta the villi increase in size and complexity to form 15 to 20 clumps of chorion bathed in maternal blood.

For the rest of the pregnancy the placenta takes over the tasks of foetal maintenance. The maternal blood entering the placenta flows into pools or sinuses where it comes into close contact with the blood vessels in the foetus. Although the blood of mother and foetus never actually mix, the foetus can obtain nutrients and oxygen from the mother and get rid of carbon dioxide and other wastes. Antibodies from the mother enter the foetal circulation to provide protection from diseases such as diphtheria, smallpox and measles.

Despite the intimate connection between foetal and maternal tissues the foetus is essentially liberated from the mother when the ovum is shed from the ovary, for, although retained in the uterus, it is no longer part of the mother's tissue. And the foetus does control its own destiny in one sense, for the placenta makes the hormone progesterone, responsible for maintaining the pregnancy, plus the hormone estrogen which, at the end of pregnancy, helps to initiate labour and thus birth.

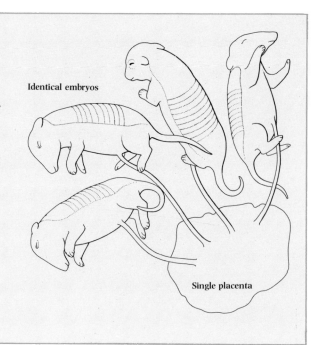

IDENTICAL BABIES
Many mammals produce more than one baby at a time. This is usually because the ovaries produce a number of eggs simultaneously which are all fertilized. Occasionally in humans, and in cows, sheep and pigs among others, a single fertilized egg gives rise to more than one embryo. The offspring all have exactly the same genetic complement, as they come from divisions of a single cell. The armadillo regularly gives birth to 4 identical embryos which share a single placenta.

Identical embryos

Single placenta

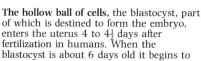

Blastocyst

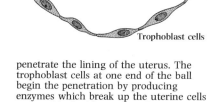

Uterine lining

Trophoblast cells

The hollow ball of cells, the blastocyst, part of which is destined to form the embryo, enters the uterus 4 to 4½ days after fertilization in humans. When the blastocyst is about 6 days old it begins to

penetrate the lining of the uterus. The trophoblast cells at one end of the ball begin the penetration by producing enzymes which break up the uterine cells and enable the blastocyst to burrow in.

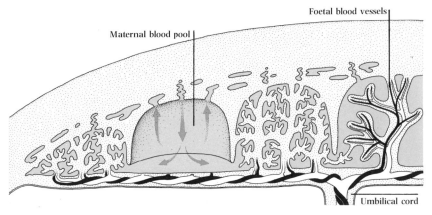

Foetal blood vessels

Maternal blood pool

Umbilical cord

A mature placenta is made up of some 15 to 20 close-packed lobes or cotyledons. Each lobe is the site of vital nutrient and gaseous exchanges between the mother and her developing child. Each cotyledon

contains treelike bunches of foetal blood vessels which protrude into maternal blood pools (intervillous spaces). Arteries of the mother's circulation bring oxygenated blood to the pools. The blood drains back

In less advanced mammals the 2 reproductive ducts (the oviducts and the uteri) remain separate for most of their length. In some placental mammals, such as mice and rabbits, there is a single vagina but 2 long, separate uterine arms. Each elongate uterus can support many embryos at once, each with its own placenta. An adaptation is found in the impala, *Aepyceros melampus*. Here only the right arm of the uterus develops properly and, although both ovaries are active, blastocysts always implant in the developed arm.

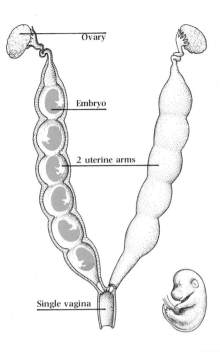

Ovary
Embryo
2 uterine arms
Single vagina

The ultimate development of the placental mammal's reproductive system is seen in primates, including woman, some bats and armadillos. The Fallopian tubes from each ovary open directly into a single central uterus, formed from the fusion of the 2 lateral uterine arms. Usually only a single blastocyst implants at a time in this simple uterus. Rarely there is an abnormal development of a human female's uterus in which the 2-armed structure remains and the uterus is divided down the middle by a wall of tissue.

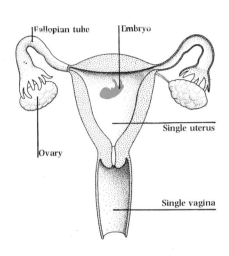

Fallopian tube
Embryo
Single uterus
Ovary
Single vagina

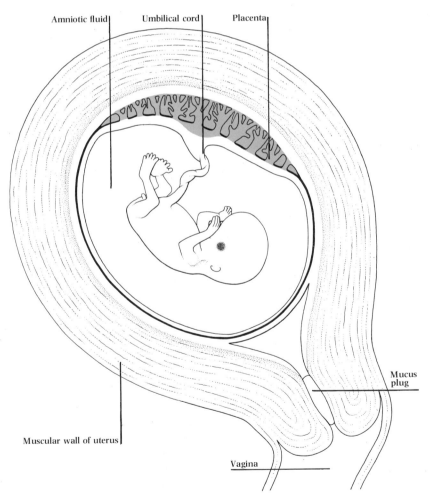

Amniotic fluid
Umbilical cord
Placenta
Mucus plug
Muscular wall of uterus
Vagina

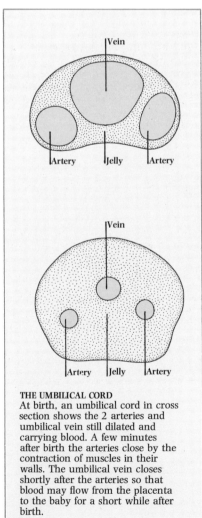

Vein
Artery
Jelly
Artery

Vein
Artery
Jelly
Artery

THE UMBILICAL CORD
At birth, an umbilical cord in cross section shows the 2 arteries and umbilical vein still dilated and carrying blood. A few minutes after birth the arteries close by the contraction of muscles in their walls. The umbilical vein closes shortly after the arteries so that blood may flow from the placenta to the baby for a short while after birth.

into the maternal circulation through the many veins which open all around each lobe. Normally the pools contain about 150 cc of blood, which is replaced 3 to 4 times a minute. Oxygen, nutrients and antibodies pass into the foetal blood, and carbon dioxide and waste products pass from the foetal to the maternal circulation. Amniotic fluid cushions the foetus from external traumas.

Entering the world

At the moment of birth, a mammal is traumatically thrust into the world to begin life as an independent individual. The events of birth are complex, their timing and synchronization carefully controlled by many different hormones.

Because the whole object of the birth process is the delivery of an appropriately developed foetus, it is not surprising that foetal hormones play a vital role in the initiation of labour. During pregnancy, the muscles of the mother's uterus increase in size and strength. They may contract occasionally and irregularly, but the really large rhythmic contractions characteristic of labour are prevented by the hormone progesterone while the foetus is developing. At first this hormone is made by the ovary, but later in pregnancy it is manufactured by the placenta.

At the end of its maturation the foetus signals that it is ready to be born. What probably happens is that the hypothalamus in the brain of the foetus becomes able to stimulate the pituitary gland to produce the hormone corticotrophin. In turn, corticotrophin stimulates the synthesis and release of hormones from the foetal adrenal glands which lie over each kidney. These adrenal hormones promote the conversion of the progesterone released by the placenta into estrogen which works in exactly the opposite way to progesterone—it stimulates the muscles of the mother's uterus to contract.

The hormones from the foetal adrenal glands have other effects too. They cause the placenta to produce a chemical called prostaglandin F2 which is a powerful stimulant to uterine contractions. Further uterine stimulation may also be exerted by the pressure of the foetus on the exit to the uterus, for this is thought to trigger manufacture of the hormone oxytocin from the mother's pituitary. By all these means the contractions build up until finally the infant is pushed out into the world.

Now a new chapter opens as the young animal has to fend for itself. The lungs, collapsed before birth, must very quickly be expanded and filled with air. To aid this, the lung lining produces a substance called surfactant, which lowers surface tension. The blood circulation must also change. Blood flow from the placenta must be closed down and flow to the lungs opened up. In this complex procedure timing is literally vital. The young mammal must prepare for a drastic change in diet. Formerly sustained on small molecules of predigested food supplied directly to its blood via the placenta, it must now digest food for itself. Its gut now has to release the chemicals essential to break down the fats, complex sugars and proteins of the mother's milk so that they can be absorbed into its blood circulation.

BIRTH OF A DOLPHIN
Whales and dolphins usually give birth to a single large young which emerges tail first. In a whale foetus the head and neck are heavy and rigid and the tail light and mobile, so a tail presentation position is naturally assumed toward the end of gestation. The problem of breathing before birth is complete does not apply. Whales are born underwater and must be completely liberated from the mother's body before they can surface and breathe. Fins are folded back against the body so they do not impede progress through the birth canal. Once the baby is born, the mother, often helped by other females, pushes it to the surface to take its first breath.

At the moment of birth vital changes take place in the baby's circulation. The blood of a foetus is oxygenated in the placenta, not the lungs. At birth the lungs must take over. In foetal circulation oxygenated blood flows from the placenta to the heart's right atrium via the umbilical vein. Most of this blood is swept into the left side of the heart through a hole in the heart, the oval foramen. It is then pumped to foetal tissues and organs. Lung circulation is by-passed via a special foetal feature, the ductus arteriosus. During birth the lungs become inflated and the ductus arteriosus closes by contraction of its muscular walls. This forces blood from the right ventricle into the lung circulation. Blood return from the lungs increases pressure in the left atrium and closes a valve over the hole in the heart. The umbilical arteries close, and then the umbilical vein.

Primates and ungulates generally give birth to one large baby at a time. The birth itself may take some time, and it is important for the baby to emerge head first so that it may breathe before the process is complete. If the head was in the uterus at the moment of the first breath, the baby might choke or become infected with non-sterile amniotic fluid. The rangy forelegs of an ungulate, such as a zebra, also appear with the head, hoofs forward. In animals producing a litter of small young, some appear head first, others tail first.

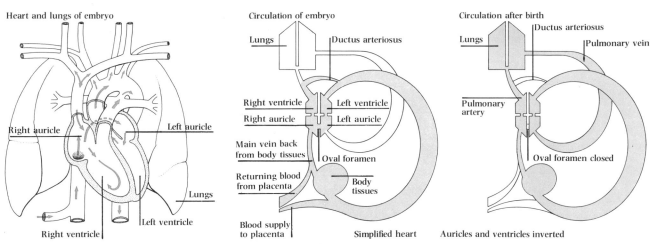

Heart and lungs of embryo

Right auricle

Left auricle

Lungs

Right ventricle

Left ventricle

Circulation of embryo

Lungs

Ductus arteriosus

Right ventricle

Left ventricle

Right auricle

Left auricle

Main vein back
from body tissues

Oval foramen

Returning blood
from placenta

Body
tissues

Blood supply
to placenta

Simplified heart

Circulation after birth

Lungs

Ductus arteriosus

Pulmonary vein

Pulmonary
artery

Oval foramen closed

Auricles and ventricles inverted

Once their young are born, after a period of incubating eggs or a pregnancy, animal parents must care for, protect and feed their offspring. Not all animals do care for their young of course; many invertebrates leave the survival of the eggs to chance, although the social insects, termites, bees and ants, look after the eggs and larvae of the colony with great efficiency.

A few birds, the mallee fowl for example, have no contact with their young after they hatch; ducklings and chickens are almost immediately independent and can move about and find food for themselves. But for the majority of birds with helpless young, the parents' real work is just beginning when the eggs hatch. As soon as they emerge from their eggs the young birds begin clamouring for food. The task of constantly feeding the fast-growing and ravenous chicks is so demanding that both parents are often needed to maintain an adequate supply of food. Hence most birds are monogamous, often pairing for life. If they live in a particularly rich habitat with plenty of food, then it may be just possible for the female to raise the young alone. Males then become polygamous, mating with a succession of females, but leaving them to bring up the young alone.

Birds that care for their young in a nest must also keep the nest clean. Parents remove faeces by swallowing them or carrying them away, or the chicks eject faeces over the edge of the nest. Parents keep their young warm with their own bodies and will shelter them from sun or rain with their wings.

In addition to providing food, warmth and shelter, parents protect the chicks from the constant threat of predators. Colonial sea birds attack predators together, dive-bombing them from the air. Others carry young away from danger or give warning calls to alert them; plovers distract a predator and lure it away by feigning injury.

Most bird behaviour is not the result of active teaching by parents. Chaffinches, and other songbirds, inherit a rough version of their song and perfect it by hearing adults sing.

A few birds find the task of rearing young so daunting

that they attempt to divest themselves of their responsibility altogether. Brood parasitism involves parent birds tricking other birds into becoming foster parents for them. European cuckoos lay matching eggs in the nests of other species such as warblers. The young parasite hatches out first and promptly tips out the host's eggs. He then receives all the food from his doting foster parents. They continue to feed him as he grows into a bird many times their size. This task is so taxing that the foster parents often lose weight themselves and may even die of the strain. Meanwhile, the adult cuckoos fly back to Africa as early as July, leaving other species patiently bringing up young cuckoos for the rest of the summer. Brood parasitism in various forms is found among a number of different bird families. Widow bird parasites mimic the host young completely and are brought up undetected with the rest of the family. The young honeyguide parasite kills its nest companions.

Mammals are not confronted by the same problems as birds. They have taken the care of their young to its logical conclusion; the mother retains them inside her body for weeks or even months. When she gives birth, it is to a large, well-developed individual which may be up to a third of her own size. But even after birth, the young mammal requires a prolonged period of care which may last for years in carnivores and primates. In the early stages there are no feeding problems, as the mother provides food in the form of milk, which is produced in her own mammary glands. These glands are believed to have evolved from sweat glands; the duck-billed platypus still feeds her young milk secreted by more than a hundred tiny glands which open on to the skin of her abdomen. Milk is a highly nutritious substance containing fat, protein and carbohydrates as well as special antibodies to protect the young from infection. Pigeons also produce a form of milk from their crop for feeding the young. As in mammals, its production is controlled by the hormone prolactin.

Like birds, mammals not only feed their young, but also protect them through the vulnerable early stages of life. Even though mammals are warm-blooded, some young are born naked and thus need to be protected against the cold. Rabbits, badgers, beavers and many others build nests of some form and stay with their young to provide warmth from their own bodies. Since predators are a constant threat, many mammals hide their young in sheltered places or dens. If danger is persistent, they pick the young up in their mouths and carry them to another, safer hiding place. Adults are always on the lookout for danger, and usually warn each other and their young by signals which cause them to scatter or run away fast. Among common visual signals are the flashing of a white tail or the curious jumping movements of some gazelles. Fortunately, the young of many herd animals are born in an advanced stage of development; although shaky on their long legs, they can stand and run a few minutes after birth. Keeping close to their mother and up with the herd can literally mean the difference between life and death. Sometimes, however, a young one is cornered; then it is not unusual to see a mother bravely attempt to defend her young. Normally docile animals such as wildebeest can suddenly become aggressive when they have young to protect, as many an unwary hunter or farmer can testify. The drive to protect the young, and so ensure the continued survival of an individual's own genes, is an extremely powerful and necessary force in evolution. In a biological sense, a female has invested her future in her offspring, and so it is not surprising that she will go to almost any lengths to ensure their survival.

Many young mammals spend much of their more relaxed moments at play. Play appears at first sight to be nothing more than a pleasurable way of spending time. But, in fact, during play, the important tasks of adult-hood, such as fighting and hunting, are being learned. By constantly repeating such activities, the young animals are actually practising and improving skills. When a real fight or hunt does take place, there will be no time for practice, and one mistake may prove fatal.

Play provides the young with opportunities to make mistakes when they will have no dire consequences. It strengthens the muscles and sharpens the reflexes of the animal's developing body. Young mammals, carnivores, primates and whales in particular, have an insatiable curiosity to explore, investigate and play with any moving object. The confidence to approach and contact will help them learn rapidly about their environment and how to survive in it. Young monkeys are notorious for their boisterous, romping play behaviour; they constantly chase, wrestle and fight with each other. The time spent with their mother was thought to be the most critical for normal development, but recent research appears to disprove this theory. Monkeys, like all primates, are social creatures and must learn to communicate and live with each other in families or larger societies. The most effective way of learning how to integrate is to mix freely with lots of other individuals. Similarly, educationalists and child psychologists, as well as most parents, now recognize the importance of early learning in humans, and favour crèches and pre-school playgroups for young children.

Formerly, much animal behaviour was assumed to be largely or wholly inherited or 'instinctive', but it is becoming increasingly apparent that learning and experience play a vital role in its development. Many invertebrate animals live only a short time and their limited behaviour patterns are frequently inherited and 'built in' to the nervous system. Even in vertebrate animals, some behaviour, such as finding the right place to live or recognizing a mate, still appears to be largely genetically controlled. Nevertheless, learning plays a vital role in the behaviour of higher animals. Learning brings with it the opportunity to acquire new and favourable types of behaviour. By their ability to learn from mistakes and to exploit new situations, higher animals are flexible and able to adapt to changes.

The importance of early learning during development, whether through play, exploration, trial and error, or processes such as imprinting, is only just beginning to be appreciated by scientists. It is now obvious that the young need not only food, warmth and protection from their parents, but also the opportunity to learn about their environment and the other living creatures with which they will share it.

CARE AND LEARNING

A cub pouncing on a fallen leaf is not merely skittish but gaining essential experience for the future—one day a live prey will take the place of that leaf.

Mother's milk

By suckling their young, mammal mothers provide their offspring with food in the form of milk, and extend the period of parental care long after birth. During evolution, the milk-producing breasts or mammary glands may well have first functioned as modified sweat glands which became active during a breeding season to help keep the young moist. New-born creatures probably licked the glands for a drink. Eventually the glands became modified so that the secretions contained food substances. The milk glands of the duck-billed platypus seem to represent an early stage in breast evolution. On the underside of the mother, a hundred or so small glands open separately on to the skin surface. She has no nipples and the milk simply oozes out on to special hairs from which it is licked up by the young.

The milk glands of placental mammals are much more specialized. The glands develop at intervals along the milk lines—a pair of skin ridges running the whole length of the body underside. Pigs and dogs, which give birth to large litters, have many nipples along the entire length of each milk line, while primates, elephants and horses, with small numbers of young, have one pair.

Each active, developed mammary gland is rather like a bunch of grapes. Every 'grape' or alveolus is lined with cells that produce milk and release it into a central collecting space. From here milk flows into channels, which lead to the surface, usually via a nipple.

Milk production does not begin until a hormone—prolactin—is made by the mother's pituitary gland after the birth of the young. At this time the levels of the 'pregnancy' hormones progesterone and estrogen fall suddenly, and this stimulates prolactin release. The continued flow of milk depends, however, on the stimulus provided by the sucking of the young which stimulates prolactin and also triggers the release of another pituitary hormone, oxytocin, which causes the muscles of the mammary gland to contract and eject milk. In whales, because the pups do not have mobile lips for sucking, specially developed muscles squirt milk into the mouths of the young. When the sucking stimulus stops at weaning, milk production also ceases.

Milk is a highly nutritious fluid rich in fats, proteins (particularly easily digested casein) and milk sugar or lactose. It also contains minerals, vitamins and antibodies which help to protect the young against disease. The precise composition of mammalian milk varies with the needs of the young. The fat and protein content tends to increase while the sugar content decreases during the process of lactation.

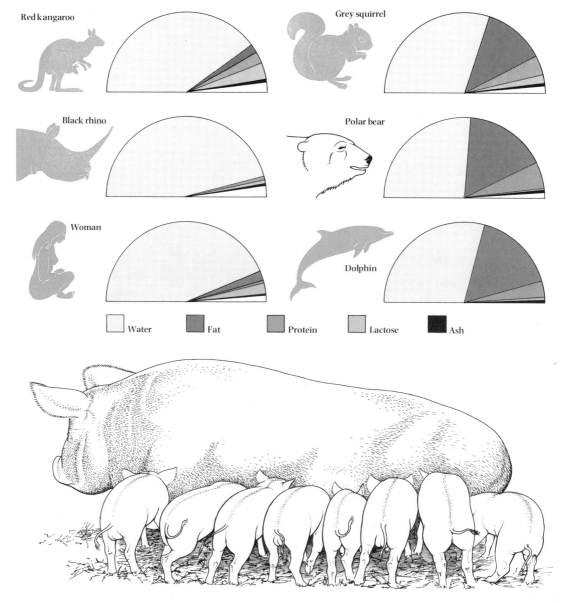

The composition of milk seems to be linked to an animal's habitat and to the needs of the sucking young. Mammals living in arid conditions produce milk with a high water content. Marine and arctic mammals have milk with a high fat content—about 33 per cent in the polar bear and dolphin—to supply the massive energy needs of young which must grow rapidly in a harsh environment. Milk can change during lactation. The milk of a red kangaroo for a new baby contains little fat, but milk for advanced young contains up to 20 per cent fat.

Red kangaroo

Grey squirrel

Black rhino

Polar bear

Woman

Dolphin

☐ Water ▨ Fat ▨ Protein ▨ Lactose ■ Ash

Mammary glands develop along 2 milk lines which run down most of the torso. In animals such as pigs, which produce large numbers of young, a series of glands are present along the length of the mother's milk lines. The number of mammary glands is linked to the number of young generally produced in a litter. Pigs have up to 6 pairs of nipples and 12 piglets in a litter.

Before a first pregnancy a breast is composed mainly of fat tissue. During pregnancy, stimulated by high levels of the hormones estrogen and progesterone, a complex branching duct system develops. Each duct ends in a pocket, alveolus, of milk-secreting cells, and these are surrounded by muscle cells. Groups of alveoli form lobules and a number of lobules form a lobe. There are ducts from the lobes to the nipple, but beneath the nipple each duct widens into a pouch for milk collection and storage.

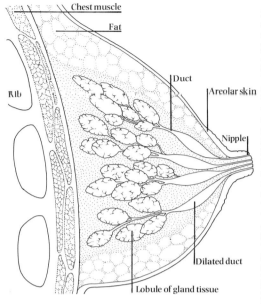

Chest muscle
Fat
Rib
Duct
Areolar skin
Nipple
Dilated duct
Lobule of gland tissue

Nipples and teats are different in structure. Nipples of primates and humans are elevations into which all the main drainage channels of the breast open, so the baby sucks milk directly from the mammary ducts. A cow, however, has milk reservoirs in the form of elongated teats. Milk ducts open into the cavity of the teat which can store up to $\frac{3}{4}$pt (0.42 1), and a single duct, the streak canal, drains the teat. Thus when a calf sucks, it receives milk that has been secreted into and then perhaps stored in the teat.

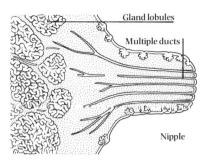

Gland lobules
Multiple ducts
Nipple

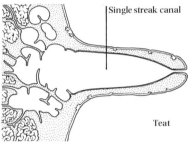

Single streak canal
Teat

The female gorilla, *Gorilla gorilla*, usually produces one baby at each pregnancy. She has a pair of mammary glands which develop on the milk lines in the chest region.

Starting to feed

Feeding young birds is hard work for the parents. Eagles and hawks must bring a couple of large prey to the nest each day, and small, insect-eating birds have been observed making 30 to 60 feeding trips each hour, amounting to thousands of trips during the period of care.

But not all birds provide food for their young. Young birds that are well-developed and highly mobile soon after hatching, such as ducks, plovers, hens and megapodes, are ready to start pecking and searching for food right away. They discover what is edible by trial and error and by following and watching their mother. Some pheasants and chickens may help their young to find food by pointing to it with their beaks. Others, such as rails, may feed the young.

Passerines or perching birds, hawks, owls, woodpeckers, pelicans and many others, hatch out young which are blind, naked and helpless. The fledgelings can do little more than gape for food and depend entirely on their parents for feeding while in the nest. Even after they have left it, they may still be fed for many weeks until they become totally independent.

The number of eggs a female lays is thought to be controlled by the number of young the parents are capable of feeding. In general, birds with helpless young are monogamous, as both parents are needed to cope with feeding. Only when food is plentiful does a male become polygamous, leaving the females to raise the young themselves.

Normally the eggs are laid soon after one another, and are synchronized to hatch out at the same time so that the young are of the same age and size. Owls, on the other hand, lay their eggs several days apart, and thus their young are frequently of varying ages and sizes. When food is in short supply, competition is keen, and the smallest and weakest owlets are likely to starve, as their older brothers and sisters monopolize the food. The golden eagle lays only two eggs several days apart. But the elder of the two chicks invariably eats the younger to ensure that he gets all the food for himself.

Since young birds grow extremely fast, they must make sure that they have enough food to survive. To do this, they may deprive their siblings of their share, or even eat them. The young birds also have a number of adaptations that are designed to stimulate their parents to feed them. As soon as a parent lands on the nest, the chicks stretch up and open their mouths wide. This gape serves as a stimulus to release the feeding reflex in the parent.

As the adult robin lands on the rim of the nest, the young bird stretches and gapes so that the parent can simply pop the food straight into its waiting mouth.

The chicken does not give its young much help, but will point with its beak at suitable food to encourage the young bird to try it for itself.

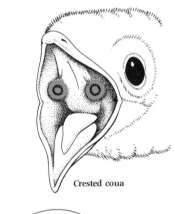

Crested coua

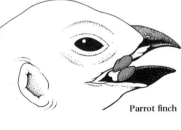

Parrot finch

A parent bird is often attracted and stimulated to feed its young by seeing the bright markings which many nestlings have in or around their mouths. In some birds the light-coloured margin of the mouth is simply enlarged to attract attention, but others have more elaborate systems. The parrot finch has pearly reflection nodules at the corners of its beak which shine out from the dark nest. Others, such as the crested coua, have markings on the palate or on the tongue.

THE DISCUS FISH
Mammals and birds are not the only animals capable of feeding young on their own body secretions. The extraordinary discus fish, *Symphysodon discus*, allows young to feed on the mucus secretion covering its body. Deprived of this highly nourishing food the newly hatched young die, even though plenty of other food is available. Both parents share feeding duties.

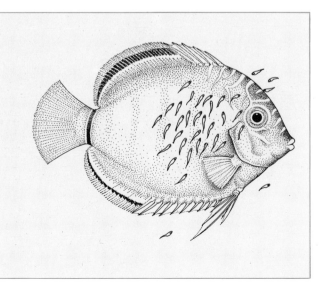

Young pelicans thrust their heads deep into the parent's large gullet and help themselves to any half-digested fish they can find there.

Birds of prey, such as hawks and falcons, bring back large prey to the nest and feed their young by tearing off pieces of flesh of a size that the young can swallow.

Gulls often catch and eat fish many miles from the breeding colony. When they return they regurgitate portions of fish for the young birds to eat.

Pigeons produce their own milk to feed their young. The lining of the crop, a sac usually used for food storage, produces a thick secretion similar in composition to mammalian milk. The nestlings are fed on this for the first 4 days, and then eat grain or other solid food mixed with the milk. Crop milk is rich in protein and fat and gives the grain-eating pigeon a nourishing diet in the vital early stages of life.

The crop is an expansion of the oesophagus and, in pigeons, is divided into 2 chambers. Crop milk production is controlled by the hormone prolactin—the same hormone that controls the flow of mammalian milk.

The vigilant parents

Female mammals begin the care of their young by retaining them within their bodies, nourishing and protecting them. They then give birth to relatively large individuals which vary in their degree of development, and which require lengthy postnatal care. Marsupials have shorter pregnancies, but their tiny young usually gain equivalent protection by hanging on to the parent or living inside a special pouch for several months to continue their growth.

Although mammals are warm-blooded, some are born without a thick coat of fur and so the mother must help them maintain their body temperature. Rodents build a nest for the young over which the mother lies to provide warmth, food and shelter. If a young pup wanders off or gets lost, it gives a high frequency call which the mother can hear. She immediately retrieves the pup by picking it up by the scruff of its neck and carrying it back to the safety of the nest.

The act of carrying young in the mouth is common among mammals, as there is little chance of the mother damaging the loose hairy skin. In times of danger, dogs and cats often carry young to a safe hiding place. This behaviour is also used as a form of punishment to control unruly cubs, and as a means of showing rank within a pack. The dominant individual will bite or shake a subordinate member in a similar fashion.

If the young are to survive, particularly during the early stages, parents must also protect them against the constant danger of predators. Young ungulates generally live close to their mother in herds, which, to a certain extent, offer safety in numbers. Experienced predators, such as lions, always stalk the young or weak, so it is vital for these vulnerable members to keep up with the herd. Fortunately, the young of herd species such as wildebeest and gazelles can struggle to their feet and move about within minutes of birth.

Even when the young are able to run with the herd, predators are likely to pursue the group and attempt to make a kill. Eventually a parent may be forced either to abandon its young and escape itself, or to try and defend them. Docile species, notably the antelope and the elk or moose, become surprisingly aggressive when they have young to protect, and can frequently ward off predators with their horns and hoofs.

Older and more vigilant members of the herd often use warning signals to protect the young. If given in time, these may mean that the fast-galloping prey stands a good chance of out-distancing the predator. Species such as rabbits and some deer have conspicuous white tails which serve as efficient warning signals when flashed to the others. Some of the curious 'stotting' or jumping movements of gazelles may also serve to alert the whole herd to danger.

REACTIONS TO DANGER
Moving young away from danger is one means of protection. A female hunting dog, *Lycaon pictus*, will move her young from their den to a new one if she suspects trouble. She carries the pups one at a time in her mouth. Thomson's gazelle, *Gazella thomsoni*, make a special stiff-legged jumping movement as a warning signal to one another. This action may also warn off a predator, telling him that he has been spotted and that he cannot make a surprise attack. The elaborate high jump in the 'stotting' movement may also convey that the gazelle is so fast and strong that the predator should try his luck elsewhere.

By standing their ground and attacking predators, many female mammals defend their young successfully. A pack of hyenas may attempt to attack a rhino calf. Their only hope of success is for some hyenas to distract the female while others tackle the calf. As long as the female keeps near her calf she may fend off the hyenas; a female rhino is a formidable enemy and the hyenas may decide to seek easier prey.

Hunting dog

Thomson's gazelle

2

1

3

A female black rhinoceros, *Diceros bicornis,* spends most of her time just with her calf. Breeding occurs at any season and there is no definite pair bond. Gestation lasts about 15 or 16 months and the mother suckles her calf until her next calf is born. A rhino stands about 5 ft (1.5 m) at the shoulder. Its facial characteristics are obvious, but the rhino can also extend its prehensile upper lip and use it to place food into the mouth.

Protecting the young

Protecting baby birds from the many predators eager to snatch them, is a vital task for all parent birds, and one that they perform vigorously. Young birds in the nest are a tempting target, and even after leaving its protection, the chicks are still at risk until they are capable of either escaping or defending themselves.

One way parents defend their young is to fight off the predator. As a general rule, only the large birds, such as swans, geese or birds of prey, take this course of action, but they can be quite terrifying in their menacing reactions. Terns, and other colonial sea birds, tend to gang up and dive-bomb intruders into their nesting colonies. The fulmar defends itself and its young by spitting foul-smelling stomach oil at the predator. Some adults keep their young near them in well organized broods so that they stand less

chance of being picked off in isolation. Shelducks herd large numbers of chicks into crèches, where they are guarded by a few vigilant adult birds.

If danger threatens and the adults cannot ward off an attack, they often make a tactical withdrawal along with the young. Highly mobile young can be led away, but more helpless nestlings may have to be carried to safety in the parent's beak or feet. Some water birds hitch a ride on their parent's back.

Another common ploy used by parent birds, especially when the young cannot be moved from the nest, is distraction of the predator. Distraction displays occur in various forms, but the parent usually tries to give the impression that it is injured and therefore vulnerable to the predator. The parent bird also attempts to make itself as conspicuous as possible,

unlike its camouflaged young and nest. Although this behaviour seems altruistic, the parent is generally not in any real danger and it does not let the predator get too close.

Many parent birds warn their young that danger is imminent by special alarm calls that mean 'look out—danger'. Young birds react by 'freezing' immediately where they are sitting and by remaining totally silent. If camouflaged or reasonably well hidden, there is a good chance that they will not be detected, even by a sharp-eyed predator.

Whatever strategy a parent adopts, it may be putting its own life at some risk when attempting to defend or protect the young. Basically, it is the only way that the parents can ensure that their own genes will survive and be passed on to future generations.

Parent birds will sometimes actually carry their young away from a dangerous situation to a safe place. Rails use their beaks to carry young, and woodcocks can grip a young bird between their thighs. Birds of prey, such as the red-tailed hawk, carry young in their powerful talons. Penguins keep young on top of their feet as they shuffle along.

Red-tailed hawk

Water rail

American woodcock

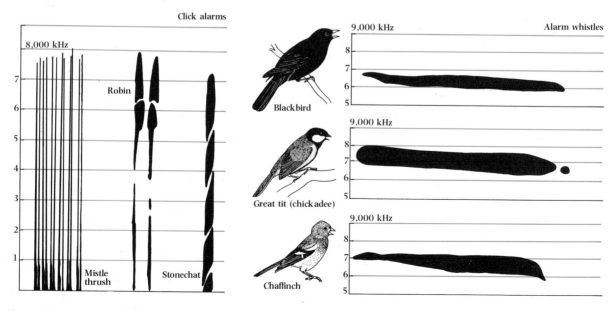

Click alarms

8,000 kHz

Robin

Mistle thrush

Stonechat

Blackbird

Great tit (chickadee)

Chaffinch

9,000 kHz — Alarm whistles

9,000 kHz

9,000 kHz

Alarm calls warn young of danger. If a hawk or dangerous predator appears, the parent gives a high, thin whistle. The whistle makes a horizontal line on a

sonogram. This type of call is difficult to locate, so the caller is in no real danger while giving it. If the bird spots a less dangerous predator, such as an owl

perched on a tree during the day, giving away its position is not serious. It makes a clicking alarm which may attract other birds to harass and mob the owl.

DISTRACTING PREDATORS

When a predator approaches the nest of a killdeer, *Charadrius vociferus*, the female makes a distraction display to protect her young. She leaves her nest, then makes herself conspicuous and attractive to the predator by feigning injury. Dragging her wing, she leads the predator away from the nest, and then suddenly flies up and escapes.

Nest

Predator

Killdeer

Large water birds such as the mute swan, *Cygnus olor*, habitually carry their young on their backs. The advantages of this are that the parent bird can move faster in the water and the young do not have to waste valuable energy trying to keep up with the adults all the time. Sitting on the parent swan's ample back also keeps cygnets warm and dry, and protects them from large, predatory fish.

Relationships

It takes a rhesus monkey—and all young primates—many years to grow up. During this time they are in constant contact with their mother as well as their social group. Initially, they depend on their mother for food, feeding from her mammary glands, and for protection. A primate cradling her newly born baby in her arms bears a strong resemblance to a human mother and her infant. This early relationship between mother and infant has been the subject of intensive research.

The mother-infant relationship was long assumed to be the most important to a young monkey's development. Controlled experiments have been carried out in which young monkeys were deprived of their mothers soon after birth and raised alone in wire cages. Although physically healthy, they grew up to be severely abnormal in other respects. The young monkeys were very aggressive and could not form social attachments or even mate with other monkeys. Another group of monkeys was raised alone in cages which contained a dummy mother to which they could cling, and from which they could even obtain milk. The substitute mother had a calming effect on the young monkeys, but they also did not develop normally. If, however, both sets of simil-

arly deprived monkeys were allowed to play with one another for a few minutes each day, they showed almost normal social and sexual behaviour.

Clearly, play with other youngsters is extremely important; perhaps even more critical than a maternal relationship. In fact, if young monkeys are raised in groups, they develop more normally than if raised solely with their own mothers. These findings support the research work of many psychologists who have studied children reared without the continued presence of their mother. Children brought up in groups on an Israeli kibbutz or a state farm in China, were found to be well adjusted, even though they were separated from their mothers for long periods.

Young monkeys and young children can learn a great deal from their parents, but it appears that the normal development of social behaviour, communication or even speech, is best achieved when they regularly interact with others of a similar age. Mother is still an important figure, but she is eventually replaced as a protector, source of food, teacher and friend by other individuals in the group. It is these interactions that will foster the independence essential for the integration of the young primate within his group.

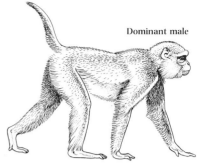

Dominant male

Low-ranking male

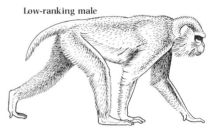

The position of a rhesus monkey's body can communicate status messages to other monkeys. A dominant or aggressive animal has a confident stance and a brisk gait. He holds his tail high and keeps his legs straight. A submissive monkey adopts the opposite signals—body held crouched and tail and head low. Contact with others must from an early stage play an important part in establishing these dominance relationships.

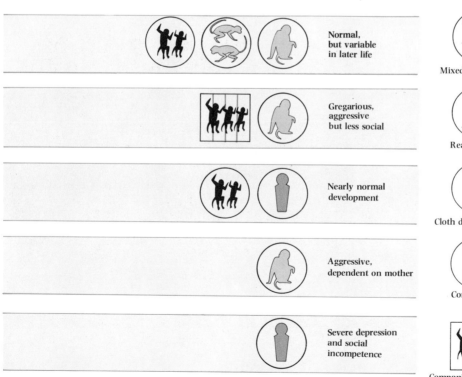

Normal, but variable in later life

Gregarious, aggressive but less social

Nearly normal development

Aggressive, dependent on mother

Severe depression and social incompetence

Mixed social setting

Real mother

Cloth dummy 'mother'

Companions

Companions in laboratory environment

Experiments to discover the relative importance of play and maternal relations to normal development have been carried out on rhesus monkeys, *Macaca mulatta*.

The monkeys were subjected to varying degrees of deprivation—raised alone, given only cloth or wire dummies as mothers, and so on. It emerged that although the

mother-infant relationship is obviously important, play with contemporaries is perhaps even more essential to normal social development.

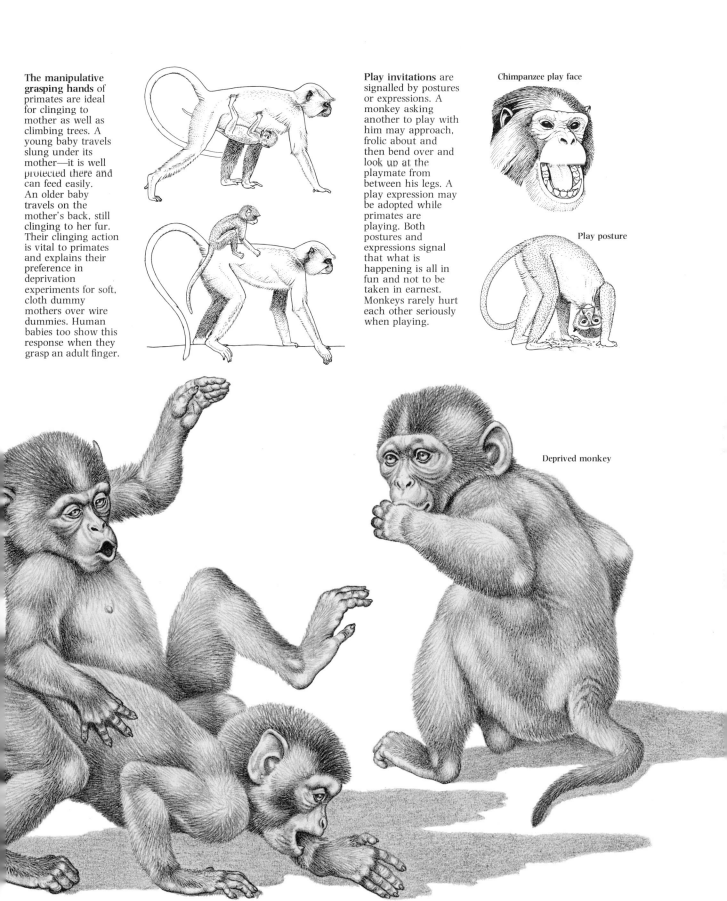

The manipulative grasping hands of primates are ideal for clinging to mother as well as climbing trees. A young baby travels slung under its mother—it is well protected there and can feed easily. An older baby travels on the mother's back, still clinging to her fur. Their clinging action is vital to primates and explains their preference in deprivation experiments for soft, cloth dummy mothers over wire dummies. Human babies too show this response when they grasp an adult finger.

Play invitations are signalled by postures or expressions. A monkey asking another to play with him may approach, frolic about and then bend over and look up at the playmate from between his legs. A play expression may be adopted while primates are playing. Both postures and expressions signal that what is happening is all in fun and not to be taken in earnest. Monkeys rarely hurt each other seriously when playing.

Chimpanzee play face

Play posture

Deprived monkey

Being social animals, rhesus monkeys must learn to communicate and relate to each other. Probably the best way to do this is through play. Young monkeys constantly chase, fight and communicate with one another, and so learn to become full members of their own special society. If the monkeys are deprived of opportunities to socialize with contemporaries they may grow up to be aggressive, non-social and incapable of forming relationships or even mating.

The meaning of play

Carnivores are fast, strong, intelligent mammals that spend a great deal of their time preying on other animals. Such a way of life would seem to be totally opposed to the carefree, relaxing world of play, yet carnivores, such as domestic cats and dogs, are among the most playful animals known. Research has shown that young kittens spend most of their early lives showing only two forms of behaviour—sleep and play —apart from feeding. It would seem that such a time- and energy-absorbing activity as play must have more significance to their lives than mere passing enjoyment.

When animals play, they do not just randomly run and jump around; certain recognizable activities occur. One of the most common is play fighting. Although a young kitten or cub constantly chases, wrestles and appears actually to fight, he always stops short of true aggression—the bites are only half-hearted nips and blood is never drawn. Nevertheless, the young animal is doing several important things. He is gaining strength by exercising muscles, he is learning what are the best ways of fighting and how far he can push his relationship with rivals in the group. Play is best seen as a form of early learning and experience, a trial and error course in which mistakes can be made without serious consequences. In later life, the young animal will often be put to the test in real fights, when a first mistake could prove fatal.

Other forms of play, such as play hunting, can be interpreted similarly. The crouching, springing and stalking of a playful kitten is essential practice if he is to become a good hunter as an adult. When cuffing and pouncing on a falling leaf, or any other moving object, the kitten is actually sharpening up his timing, reflexes and muscle power for the future hunting of prey on which his life will depend. ·

The tendency which young animals have to approach, investigate and play with novel objects has led to the saying 'curiosity killed the cat'. In fact, an intense drive to find out about and explore the environment is a major evolutionary advantage, for the young animal is, in effect, taking the initiative. As a result, the curious cat and his fellow carnivores have become some of the most intelligent and resourceful animals on earth, outwitting as well as outrunning their more timid prey. As adults they will continue to learn and sometimes play. The adult dog or cat, for example, with a surfeit of energy, occasionally will return to the childish world of play, especially in the company of his own young.

At dusk, when badgers, *Meles meles*, emerge from their sett, cubs, and even adults, indulge in energetic play. When cubs first start to come outside they are unsure in their movements, but by about 3 months old they are ready for boisterous play. They chase, try and grab each other's head or ears, roll about and even play king of the mountain. As well as helping the young to develop physically, play strengthens relationships within the group.

The average litter of a cheetah, *Acinonyx jubatus*, is about 4 cubs. A cub weighs about 12 oz (340 g) at birth and is about 26 in (66 cm) long. At 10 to 12 days its eyes are open and at 3 weeks it can walk.

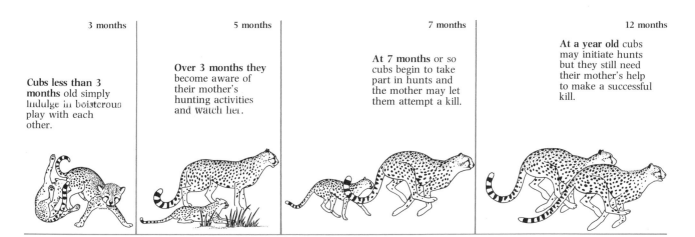

3 months	5 months	7 months	12 months

At a year old cubs may initiate hunts but they still need their mother's help to make a successful kill.

At 7 months or so cubs begin to take part in hunts and the mother may let them attempt a kill.

Over 3 months they become aware of their mother's hunting activities and watch her.

Cubs less than 3 months old simply indulge in boisterous play with each other.

A cheetah mother does not actually teach her cubs to hunt but helps them by, for example, releasing live prey for them to try out their skills. The cubs will chase and try to bring down the unfortunate prey animal. After their attempts the mother will kill the prey and they all share the carcass. Cubs become physically mature at between 13 and 16 months, and eventually the family group splits up. The cubs may not be fully adept at hunting but must abruptly start to fend for themselves.

An adult lion about to play with a cub has a particular movement to make it clear that there is no aggression intended in what is to follow. The adult lowers its forelimbs to the ground, but keeps its rear end standing; he then bobs in an unmistakable play invitation. Dogs make much the same signal but wag their tails at the same time.

Cheetah cubs indulge in rough-and-tumble play with their mothers. They nibble her ears, chase her long tail and even jump up on to her back. She is extremely tolerant of all this behaviour, which may go on for much of the cubs' first year. Gradually she channels their play and energy into the serious business of learning to hunt.

The first lessons

The behaviour patterns of invertebrates are mostly inherited from their parents. Since invertebrates generally have a short life span, it is important that the right responses be 'built in' to the nervous system. Vertebrate animals, on the other hand, live longer, and although some of their behaviour appears to be inherited, most of their responses are adapted gradually through learning and experience. Among the chief advantages of learning are greater choice and flexibility in behaviour. If, for example, a response is harmful or non-rewarding, the animal can learn to change its behaviour. Moreover, if a new situation arises, the animal has the ability to modify its behaviour as a result of its experience and perhaps exploit a new environment more effectively.

For many animals the most pressing task is to find a suitable place to live.

Habitat selection seems to be primarily an inherited behaviour pattern. Birds and mammals usually know instinctively what type of environment to occupy; even in cases where they have been reared experimentally in other than their natural surroundings, on release they tend to choose the normal habitat for their species.

Early learning plays a definite role in the need of animals of the same species to communicate with one another. Colonial sea birds, for instance, must recognize their parents or young among vast colonies comprising many thousands of individuals. Fortunately, each adult has its own unique call, which is learned by the young nestlings so that they can identify their parents by voice alone.

During their infancy many animals must also learn the distinguishing

characteristics of their own species in order to avoid choosing a mate of the wrong species later. For young birds and mammals that can walk right after hatching or birth, it is essential that they stay near their parents for food and protection. Young ducks and geese immediately follow the first moving object they see, normally their mother. But if they are hatched in an incubator and the first object they see is a human, they will become 'imprinted' upon a person instead.

Formerly, imprinting was believed to influence the later sexual behaviour of the unfortunate animal, but this is no longer generally accepted. Although some species of goose do court and display to their human foster parents for a period, most learn through experience that only those of the same species will respond to their courtship signals.

The oystercatchers, *Haematopus sp.,* feed on shellfish. With their long legs and necks they are well adapted for this type of seashore feeding. In recent years though, birds have moved inland and now breed many miles from the sea. They feed on arable farmland finding grubs and worms. They seem to have learned to exploit this different but favourable habitat to their advantage.

Seashore

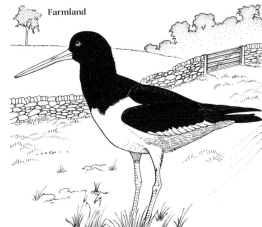

Farmland

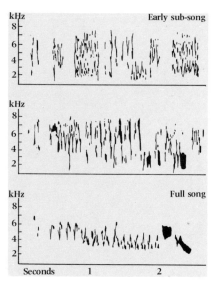

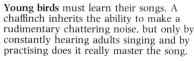

Young birds must learn their songs. A chaffinch inherits the ability to make a rudimentary chattering noise, but only by constantly hearing adults singing and by practising does it really master the song.

Baby terns in a crowded colony of birds can recognize the particular call of their parents even when they are still in the egg. When a parent bird returns to the colony where its own young are

surrounded by hundreds of other chicks, it calls from the air. Its chicks hear the call and sit up to look out for the parent. No other chicks react. This system allows efficient recognition in the vast colony.

Prairie

Woodlands

The prairie deer mouse, *Peromyscus maniculatus bairdii*, of North America, lives only in fields and avoids wooded areas. The closely related deer mouse, *P. m. gracilis*, lives only in woodlands. There are no physical differences sufficient to explain the contrast in the deer mice's choice of habitat. Even if young mice are reared in the wrong habitat they still prefer their correct habitat given the choice, indicating that the preference is inherited rather than learned. Experiments have shown, however, that early experience does seem to play a part in reinforcing the inherited preference.

Birds may have a preference for certain types of foliage habitat. A species of European tit (chickadee), the blue tit, *Parus caeruleus*, lives among broad-leafed trees, while another species, the coal tit, *Parus ater*, lives in coniferous woodlands. This preference seems to be largely inherited. If each species is hand-reared in the other's habitat and then set free, their choice of the usual habitat is still almost as frequent.

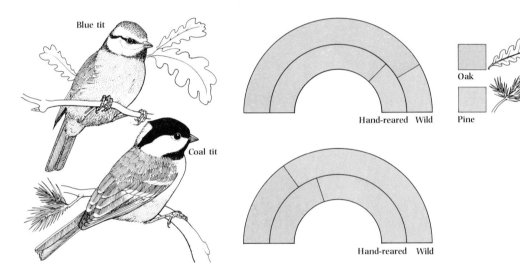

Blue tit

Coal tit

Hand-reared Wild

Hand-reared Wild

Oak

Pine

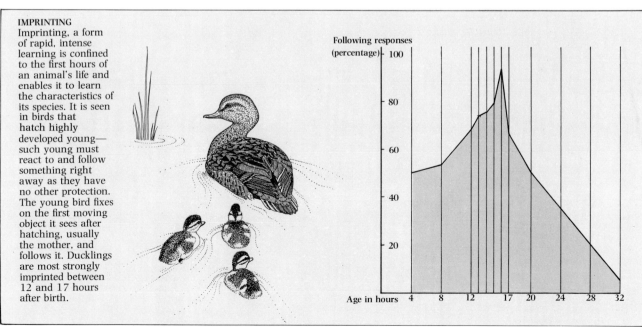

IMPRINTING
Imprinting, a form of rapid, intense learning is confined to the first hours of an animal's life and enables it to learn the characteristics of its species. It is seen in birds that hatch highly developed young—such young must react to and follow something right away as they have no other protection. The young bird fixes on the first moving object it sees after hatching, usually the mother, and follows it. Ducklings are most strongly imprinted between 12 and 17 hours after birth.

Following responses (percentage)

100

80

60

40

20

Age in hours 4   8   12   17   20   24   28   32

The foster parents

After the eggs hatch, most birds feed, protect and shelter their young—a time-consuming task. Each female lays as many eggs as possible in order to pass on more genes to the next generation. The size of a clutch is generally determined by the number of young a pair can raise. When the upper limit is reached, and there is not enough food for all, the smallest and weakest young will starve to death.

A cunning method of increasing the number of young they can produce has been developed by some bird species. After mating, they lay an egg in the nest of another species. The host species is fooled into accepting the egg as one of its own, and will hatch it and bring up the young. This behaviour, brood parasitism, requires the host birds to become foster parents.

Brood parasitism has evolved in several different avian groups: the cuckoos (Cuculidae), honeyguides (Indicatoridae), cowbirds (Icteridae), weavers (Ploceidae), and even ducks (Anatidae). The black-headed duck, *Heteronetta atricapilla*, lays its eggs in the nests of other birds.

The main problem for brood parasites is to get their own egg accepted by the host. For the European cuckoo this often involves egg mimicry—matching the size and colour pattern of the host's eggs. They must also find and parasitize a nest at the right time, so that their egg is incubated long enough for it to hatch. Most parasites lay their egg when the host clutch is still being laid. To minimize the chances of discovery further, they also remove one host egg. The cuckoo egg is timed to hatch out first, and the young parasite immediately ejects the host eggs before they hatch. As a result, the young cuckoo is able to have all the food the host parents bring, and quickly grows into an enormous bird many times the size of its host.

African honeyguides lay their eggs in the nests of hole-nesting species, such as starlings. The female honeyguide destroys the host's eggs before laying hers. But if any others are laid later, the young parasite will use special hooks on its beak to lacerate and kill the young which hatch.

Cowbirds do not destroy their host's eggs or young; they are content to take their chance of being raised with the rest of the brood. One of the most sophisticated attempts to mimic the host species is found among the weavers. The African widow birds (viduines) are parasitic on grass finches (estrildines). The young widow bird parasite resembles the young of the host brood and only adopts its adult plumage after leaving the nest.

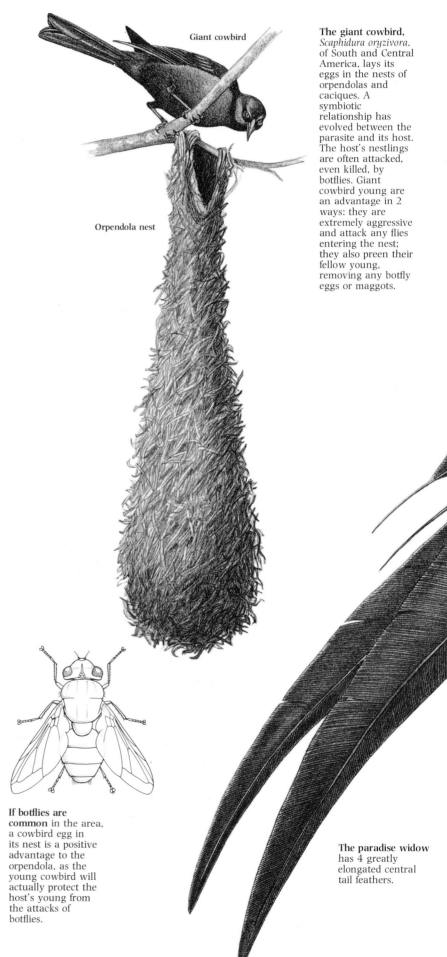

Giant cowbird

Orpendola nest

The giant cowbird, *Scaphidura oryzivora,* of South and Central America, lays its eggs in the nests of orpendolas and caciques. A symbiotic relationship has evolved between the parasite and its host. The host's nestlings are often attacked, even killed, by botflies. Giant cowbird young are an advantage in 2 ways: they are extremely aggressive and attack any flies entering the nest; they also preen their fellow young, removing any botfly eggs or maggots.

If botflies are common in the area, a cowbird egg in its nest is a positive advantage to the orpendola, as the young cowbird will actually protect the host's young from the attacks of botflies.

The paradise widow has 4 greatly elongated central tail feathers.

The exotic African widow birds use the plain estrildine finches as foster parents to incubate their eggs and care for their young. Close associations have evolved between widows and estrildines, so that each species of widow bird has become adapted to mimic the particular host it generally uses. The paradise widow, *Vidua paradisaea*, below, parasitizes the finch, *Pytilia melba*, and the adult widow can even mimic the finch's song; the young widows mimic the plumage, gape and calls of their foster brothers.

Male widow bird

Host

Parasite

The mouth lining of the young of each species of estrildine finch has a particular pattern of dark spots in a symmetrical design. A parent finch will only feed young with the correct gape markings, so widow bird young must mimic the gape pattern of their particular host. Young also mimic begging calls. Sometimes a begging posture is needed before the parent finch will regurgitate seeds, and even this is faithfully copied.

Estrildine finch

Parasite

Host

The eggs of both widow bird and finch are usually white. Whereas the plumage of the adult widow and the adult finch are utterly different, their young have almost identical plumage in dull tones of brown and grey. This resemblance is vital, as it ensures that all the young will be brought up by the foster parent. Only after the post-juvenile moult, when it no longer needs the care of its foster parent, does the widow bird drop its disguise and change dramatically to its true plumage.

133

R ather like human beings, wild animals live two sorts of lives. One is a public life, carried out in full view of anyone who cares to watch; the other is a secretive private life usually conducted within the nest or burrow. In their public lives, animals show how they catch their prey, how they defend their territories and sometimes even how they build their nests. But the details of their courtship and mating, and the way in which they care for their new-born young, frequently elude even the most persistent observer, because these activities are carried out in hidden places.

It is as well that animals should be invisible at crucial times in their life history, when their attentions are diverted from the ever-present dangers of the world outside. For if man cannot see them, the chances are that their predators cannot either. This is, without doubt, why intimate behaviour has become associated with privacy throughout the animal kingdom.

Another reason why human knowledge of the private behaviour of animals is less complete than that of their public behaviour is that man does not fully comprehend the sensory systems involved. He can, to be sure, appreciate the flamboyant visual courtship displays of peacocks and birds of paradise, but is unlikely to be able to understand fully the subtle, smell-based cues which control the courtship of tortoises, snakes and dogs.

Ethologists—scientists who study the behaviour of wild animals, usually in their natural surroundings— resort to many tricks in order to pry into the intimacies of animal behaviour. To help them, they make use of the advanced technology of the modern age. For ethologists, the greatest help by far has come from the development of cinematography and, in more recent years, of filming on to magnetic tape. The ever-decreasing size of cameras and the production of highly sensitive film has meant that permanent records can be made in the most inaccessible of places and under the very worst photographic conditions.

Apart from 'hardware', stamina, patience and inventiveness are prime requisites for wildlife photography. Hans Sielmann, one of the first ethologists to attempt to film inside hollow trees to reveal the details of the private lives of woodpeckers, was a pioneer of the modern method. To make his film, Sielmann excavated a hole on the other side of the tree into which he could insert his camera lens. A more extensive version of this technique has been used for studying the behaviour of the badger within its sett. A completely artificial sett, made of concrete, was built within the territory of a badger clan which had long been under the surveillance of the photographer. The sett was fitted with floodlights and a one-way glass screen. Once the badgers 'discovered' the sett and took up residence, the lights were turned up a fraction more each day until there was ample illumination for filming. For the first time the behaviour of the sow toward her new-born cubs could be observed. Similarly, glass boxes with cameras mounted on all sides have provided glimpses of many important types of behaviour that have long been a puzzle. The birth process in small marsupial mammals, for example, is seen to be extremely rapid and it is now certain that the mother does nothing to aid the tiny joey in its journey to the safety of the pouch.

The amateur naturalist, with little more than time and patience at his disposal, has, paradoxically, a more important part to play in the high-technology age of animal observation than ever before. Professional scientists seek to quantify their material to such an extent that the basic natural history of many species is ignored in the process. Ethologists have spent much time trying to assess whether foxes actually attack and kill sheep. They have derived means of analyzing the contents of the fox's stomach and faeces, but have relied on conversations with farmers and gamekeepers for evidence that the fox actually kills live sheep rather than scavenging on dead carcasses. An English naturalist sought to resolve this problem by dressing himself in pieces of sheepskin and joining a flock on his hands and knees. After a few rather damp and smelly evenings he reported that even the boldest fox is frightened away by the threatening action of a ewe. What clearer evidence could there be of the value of observations gleaned by an intellectually resourceful amateur?

Birds are much easier to study than mammals, and a simple blind made of hessian and bamboo canes allows an observer to approach quite close to a nest. Sometimes it is necessary to fool the birds into thinking that the blind is unoccupied. During one study of the breeding of rooks, conducted from a blind mounted high in the tree tops, it became clear that the rooks were disturbed by the sight of an observer climbing up the long ladder to reach the blind. But if two people climbed up and one immediately climbed down again the birds resumed their normal activities!

In reading and digesting the contents of the previous pages the reader will have become aware of many of the problems that face scientists in their task of unravelling the complexities of animal behaviour. He may not be so aware of the research techniques that can be used to solve some of these mysteries, of the methods ethologists use, or of the tools at their disposal.

When considering family behaviour it is absolutely essential for the observer to be able to distinguish between male and female. If the species under study is one in which there is a marked difference between the sexes (sexual dimorphism) as in sage grouse, chaffinches or lions, there is no problem. But in many animals, particularly sea birds and small mammals, there is little visible difference between the sexes. In extreme cases it may only be possible to ascertain the sex of animals under scrutiny when one of them shows behaviour exclusive to the female, such as egg laying.

The role of hormones in sexual and reproductive behaviour is an ever-recurring theme. Ethologists have pinpointed the involvement of hormones by using two distinct techniques: one is to deprive an individual of its natural source of a hormone, the other is to implant or inject preparations of the hormones into deprived or intact members of the population. Such deprivation and implantation experiments have done much to teach ethologists about the physiology which lies behind typically male and typically female behaviour.

The male hormone testosterone is responsible for territorial behaviour and the attainment of high social status in male animals. As it is made in the testes, a simple operation of surgical castration deprives the individual of its testosterone supply. As a result, castrated male grouse, for example, are soon defeated by previously subordinate males and are driven from their territories. Castrated male red deer shed their antlers and loose their supremacy as harem masters. If testosterone pellets are then implanted under the skin of these animals' necks, their dominance can be restored or even enhanced. Female animals have little testosterone circulating in their bloodstream, but if a pellet containing the hormone is implanted into the neck of a female red grouse her ovaries shrink and she adopts the aggressive attitude typical of her male partners.

Apart from man, and a few other creatures, most animals have a clearly circumscribed breeding season. Monitoring of the blood hormone levels of wild animals shows that there is a gradual buildup of sex hormones in the bodies of males and females to a certain threshold. When this is reached courtship behaviour can begin. But what governs the buildup of blood hormones? In most species it is the prevailing light conditions. The light is perceived by the eyes but it is not known how the pituitary gland, which controls sex hormone production, can read the light level. But it is clear that the breeding season of most animals can be disrupted by conditions of artificial light. In Europe badgers normally give birth in January, mate immediately afterwards and give birth the next January—the fertilized embryo having been kept 'in limbo' in the uterus for about seven months. By catching newly pregnant sow badgers in February and keeping them under autumnal light conditions, they produce a second litter in July.

Disruption of the environmental conditions can also be complemented by sensory deprivation studies in which the animal is rendered deaf, dumb, blind or incapable of perceiving odour (anosomic) by surgical operation. A male golden hamster made anosomic shows no interest in a female who is at the peak of her heat and soliciting copulation immediately in front of him. The same is true of rhesus monkeys, but sexually experienced anosomic males use their memories to overcome the effects of anosomia.

Experimental techniques such as these, carried out mostly in the 1970s (and, in the United Kingdom, under strict governmental licence) are revealing the more sophisticated aspects of reproductive and family behaviour in animals. The pages that follow present a series of case histories describing the family structure and behaviour of a few animals chosen for the completeness of knowledge about them. Each is a testament to the dedication and devotion of a small group of workers, and could not be told without their observations.

FAMILY STYLES

Much of the family life of animals is carried out hidden from view, whether in a thicket of vegetation or a deep burrow. Patient and careful spying on their activities reveals the facts of this private behaviour.

The honeybee

Perhaps the best-known social insect is the honeybee, *Apis mellifera*. It lives with thousands of other individuals in a well-organized colony, either in a nest in the wild or a manmade hive. An unusual feature of its society is that all the individuals are closely related to one another, and thus constitute one large family. In addition, the society consists of castes, that is, members that are structurally different and specialized to do certain tasks. Just one female, the queen, lays eggs; the other females, which make up the vast majority of the society, are sterile workers. Only a few males, the drones, are produced to mate with queens.

When a virgin queen leaves a hive on her nuptial flight, a number of drones are always waiting somewhere outside it. Attracted by her special smell, they mate violently with her, exploding their genitalia inside her and dying afterward. The queen obtains enough sperm to last a lifetime, and never mates again. She now departs with a swarm of workers from the old hive who start to build a new nest elsewhere.

The queen lays all the eggs in one hive. To a certain extent, she controls which castes are produced. Most of the eggs are fertilized and result in females which can become either sterile workers or queens; the unfertilized ones give rise to male drones only. The queen normally keeps rival females sterile by constantly secreting 'queen substance', a pheromone produced in the mandibular glands and passed on to the workers. In spring she stops production of queen substance for a period, and the workers build a few royal, or queen, cells. The fertilized eggs laid in these will be fed with royal jelly, a protein-rich material secreted by worker bees, and eventually become new queens.

When a young worker emerges from a cell, she spends the first few days cleaning out other cells. She then helps feed the larvae pollen and honey obtained from storage cells. On the tenth day her wax-secreting glands become active and she assists in building new cells. After 20 days she leaves the hive on her first flight and, along with her fellow workers, guards the entrance against intruders. Now she is ready to become a forager, to leave the hive and bring back nectar, pollen and water. She may die within another week or two, but the colony is similar to a vast production line—tens of thousands of workers are produced in a season. Each worker's sacrifice is not in vain, for although she does not reproduce herself, her work ensures that the colony and the queen will continue producing identical copies of her own genes.

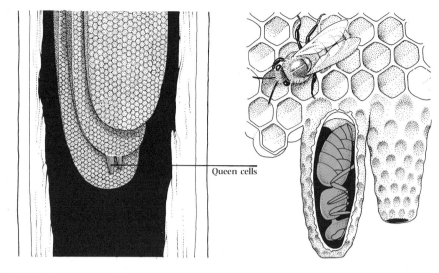

The bees' nest is a number of wax combs suspended in some form of shelter like an old tree or a manmade hive. The combs are made up of a double layer of hexagonal cells—the hexagonal shape allows all the space to be used, with no gaps between cells. Cells contain eggs or larvae, or food stores of honey or pollen. Queens are raised in long cells which hang from the lower edge of the combs.

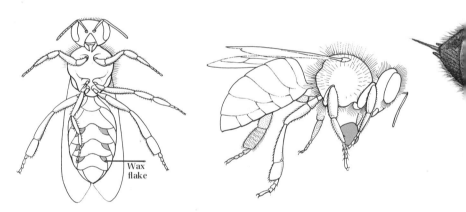

Cells are made of wax which is produced in glands on the underside of the bee's abdomen. The wax appears as thin flakes between the abdominal segments; the bee pulls the flake from the folds of abdominal skin with her middle leg and passes it up to her front legs. She kneads the wax with her mouthparts, mixing it with saliva until it is pliable enough to be used for building a cell.

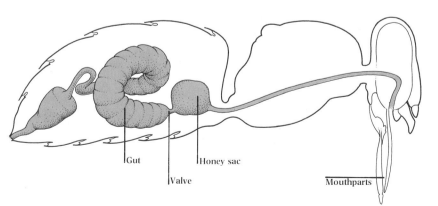

Honey is made from nectar collected from flowers. The bee sucks up nectar through the long proboscis extending from its mouth; it passes through the bee's gullet and into a storage organ, the honey sac. On returning to the nest the bee regurgitates the nectar. If the bee wants to feed on some of the nectar in its honey sac, a valve opens between the sac and the bee's stomach and nectar passes through.

There are 3 types of honeybee, each fitted to play a role in the social organization of the hive. The queen is the head of the colony and lays all the eggs. Male drones mate with virgin queens. The real work is done by the sterile, female worker bees—the majority of the colony.

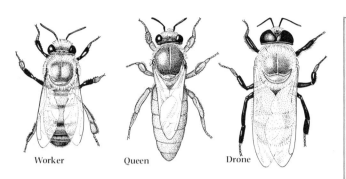

Worker   Queen   Drone

HONEYBEE FACTS

Number of bees in hive: 20,000 to 80,000 workers

Egg laying rate: efficient queen lays about 1,500 eggs in 24 hours

Development time from egg to adult:
new queen: 16 days
workers: 21 days
drones: 24 days

Life span:
queen: 4 to 5 years
workers: 6 to 8 weeks
drones: 4 to 5 weeks

Tasks of worker according to age:
nursemaid: 1 to 10 days
builder/cleaner: 10 to 20 days
forager: 20 days to death

Size of adults:
queen: 0.69 in (17.5 mm)
worker: 0.51 in (13 mm)
drone: 0.57 in (14.5 mm)

Worker bees perform all the vital tasks in the colony. They build cells and keep them clean; they tend larvae and guard the nest and, of course, they gather nectar and pollen for food stores. The worker bee has different tasks at different ages, food gathering usually occupying the second half of its life. In the summer a worker may live 6 to 8 weeks. The worker below is building up the wall of a cell.

The honeybee/2

To fill her honey sac once, a worker bee may have to visit more than a thousand different flowers. On returning to the hive, she regurgitates the nectar to a household bee who begins the process of converting it into honey, evaporating away some of the water and breaking down the sugar into more easily digested forms.

Worker bees, with their complicated structure and sophisticated array of nerve organs, are beautifully designed foraging machines. They cover hundreds of miles within a few weeks before they literally die of overwork. In addition, they are able to communicate with each other about the type, distance and direction of food sources. Although chemical communication through smell is important, visual and tactile signalling by means of curious and elaborate dances forms the basis of their language.

When a worker finds a source of food close to the hive, she collects some and returns to the hive. As soon as she lands, the other bees gather around and smell what she is carrying. She then performs the 'round dance', which seems to mean 'there is food nearby—go and look'. The watching bees become excited, and soon take off to find the food.

If the food is farther away, the forager gives more precise directions by performing the 'tail wagging' or 'waggle' dance. She runs round a figure of eight course, waggling her abdomen on the middle section of the eight. This dance provides information about the distance and the direction of the food source. The rate at which dances are performed increases when the food source is near, and decreases when it is farther away. Accurate distances up to several miles can be conveyed in this way. The direction is given by the angle that the waggle run in the centre of the figure of eight makes with the vertical plane of the honeycomb. This corresponds to the angle of the food source to the sun. The quicker the dance routine is repeated the nearer the food source. This is a sensible code because close food sources are the most useful as they can be reached with a short flying time and are therefore worth the energy spent in communicating their location.

There are, of course, additional factors that may help the foraging bees find a regular food supply. They may have learned its location, as they can learn to navigate by using local landmarks which they regularly see and pass. They may also recognize a familiar smell and associate it with a particular tree or bloom nearby. Undoubtedly, though, it is the 'dance language' which provides the most detailed information about food sources.

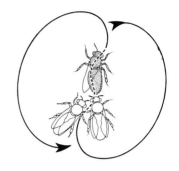

If a bee finds a good source of food more than 55 yd (50 m) from the hive, it returns to communicate the distance and direction of the food to the other bees by performing the waggle dance. The dance is in the form of a figure of eight with a straight run between the circles, when the bee waggles its body. The nearer the food the more turns of the dance are performed.

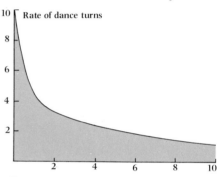

Rate of dance turns

Distance to food in kilometres

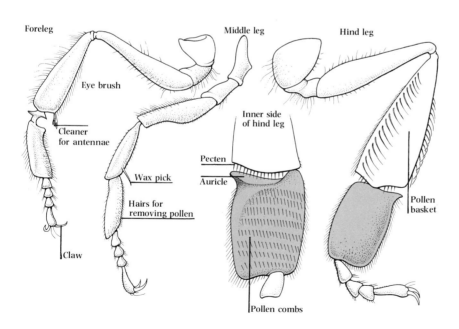

Foreleg
Middle leg
Hind leg
Eye brush
Cleaner for antennae
Wax pick
Hairs for removing pollen
Claw
Inner side of hind leg
Pecten
Auricle
Pollen combs
Pollen basket

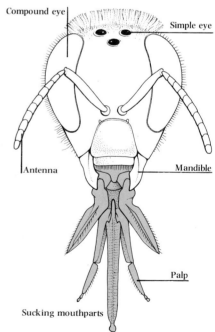

Compound eye
Simple eye
Antenna
Mandible
Palp
Sucking mouthparts

All the tools needed by a bee for its tasks are on its own body. On each foreleg is a fringe of hairs used to clean the large eyes; long hairs to remove pollen from its body, and a notch for cleaning the antennae. The middle legs have a pollen brush for removing pollen from the forelegs, and a spike for taking wax from the abdominal wax glands. Each hind leg has rows of pollen combs for removing pollen from the middle legs and body; a stiff comb, the pecten, for taking pollen from the opposite hind leg; and a rammer, the auricle, for transferring pollen to the pollen basket on the outside of the hind leg. The extended proboscis on the mouthparts enables the bee to suck up nectar even from tubular flowers.

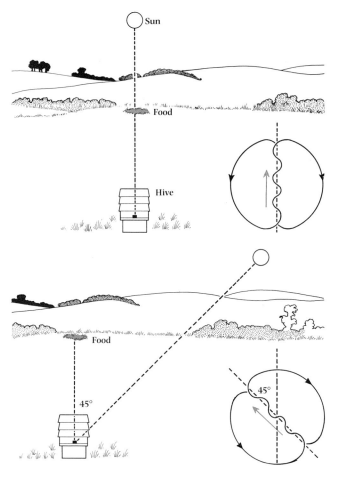

The direction of the food source is conveyed to the other bees by the angle of the straight run in the middle of the waggle dance. Having found and collected food, the bee flies straight back to the hive, noting the position of the sun on the way. Once inside the hive it performs the dance so that the angle of the straight run to the vertical corresponds to the angle of the food source to the sun.

If the food source was exactly in the direction of the sun, the bee keeps the straight run of the dance vertical. If the food source is at an angle of 45 degrees to the sun, the straight run of the dance will be at an angle of 45 degrees in the right direction. The tail-wagging movements attract attention to this part of the dance. Once the dance is over, the other bees fly off in the direction indicated.

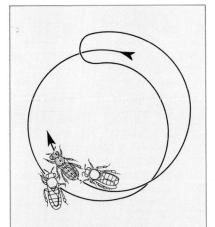

THE ROUND DANCE
When a bee returns to the hive from a food source nearby—less than 55 yds (50 m) from the hive—it performs the round dance to communicate its find to the other bees. The bee moves in small circles first to the right and then to the left. Other worker bees get excited and join in the dance. As they get close to the first bee in the dance, they pick up the scent of the flowers it has visited. The richer the food source the more energetic the dance and the longer it lasts. Some round dances are performed for only a few seconds, others for as long as a minute.

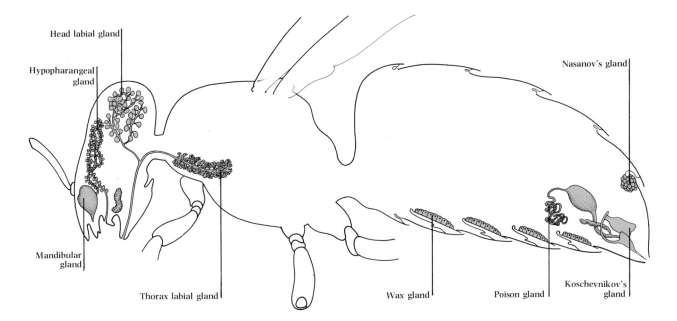

The many glands of the honeybee play a part in the social organization of the colony. The mandibular glands in the queen bee produce 'queen substance', a pheromone which maintains the cohesion of the hive. In the spring when swarming occurs, she stops making this substance. Its absence tells the workers to start to produce new queens by building special royal cells for fertilized eggs and feeding the larvae royal jelly—the food of queens. The hypopharangeal gland of the worker produces royal jelly. The Koschevnikov's gland and Nasanov's gland in the queen operate when, after the mating flight, the new queen must lead away some of the swarm for her colony. The secretions of these glands act to attract bees to her.

The salmon

Most fish spend their entire lives in either fresh or salt water, making short journeys in search of food or breeding sites. The salmon has evolved the means of spending the first, and sometimes the last, part of its life in fresh water and the remainder in the sea. Even more remarkable is the fact that the marine feeding grounds may be more than 2,175 miles (3,500 kilometres) from the freshwater breeding grounds.

At the headstream of a river, salmon choose gravel-bottomed rivulets as egg-laying sites where they make V-shaped troughs to deposit the fertilized eggs. The eggs quickly hatch into alevins, whose development varies with salmon species. In the Atlantic salmon, *Salmo salar*, the alevins grow slowly, first into parr with distinguishing dark bands on their bodies, then into silvery smolts. These smolts are two years old before they start their seaward migration, but in the Pacific pink salmon, *Oncorhyncus gorbuscha*, the fish develop more rapidly, and smolts start their descent to the sea within only a few months of hatching.

Once in the sea, the young salmon head for the edge of the continental shelf. Here the upwellings of the ocean currents are rich in nutrients, and provide ideal conditions for the growth of huge shoals of tiny shrimplike crustaceans on which the salmon gorge and which give their flesh its pink colour.

To spawn, Atlantic salmon may return to fresh water—and to the very headstream in which they were born—two or three times, but Pacific salmon usually make the journey only once. How salmon navigate precisely enough to return to within a few feet of the site where they themselves were as eggs is a mystery. The final stage, the selection of the home stream, is controlled by the salmon's excellent sense of smell, but this does not explain their vast transoceanic trek.

During the final phase of the return journey the male salmon loses his silvery appearance and becomes quite dark. In most species the jaws become pincerlike and ineffective for feeding, and in the pink salmon a grotesque hump also develops on the male's back. These males are attractive to ripe females, and a short courtship ensues in which either partner digs the trough for the eggs. About 1,500 eggs are laid and fertilized by the creamy sperm-containing milt shed over them by the male. The exhausted adults are now no match for the bald eagles and brown bears that descend on the breeding streams in search of an easy meal. Few salmon survive to repeat this extraordinary life cycle.

The salmon lays its eggs in a shallow V-shaped trough dug in the gravel of the stream bed. To make the nest, the fish swims rapidly, vibrating its tail. Its pelvic fins disturb the gravel, which is then spread to either side by the salmon's tail. The nest may be up to 6 in (15 cm) long and 0.4 in (10 mm) deep. If the migratory journey has been hard, the salmon may have only just enough energy left to complete this task.

The female pink salmon does not feed while she is in fresh water and by the time she reaches the home stream she is little more than a bag of eggs. After a brief courtship she digs her nest. With a few gentle wriggling movements she releases her eggs into the trench and covers them with gravel.

Outlet of nostril   Eye

Inlet of nostril

A salmon retains an image of the odour of its home stream in its memory from the moment it emerges from the egg. Like all fish, salmon have sensitive noses. The olfactory chambers, just in front of the eyes, are rosettes of cartilaginous flaps covered in the nasal membranes. They are shielded by a flap of skin with distinct inlet and outlet pores. Rows of tiny hairs, cilia, interspersed with the sensory membrane, maintain a constant flow of water into the chamber and over the membrane.

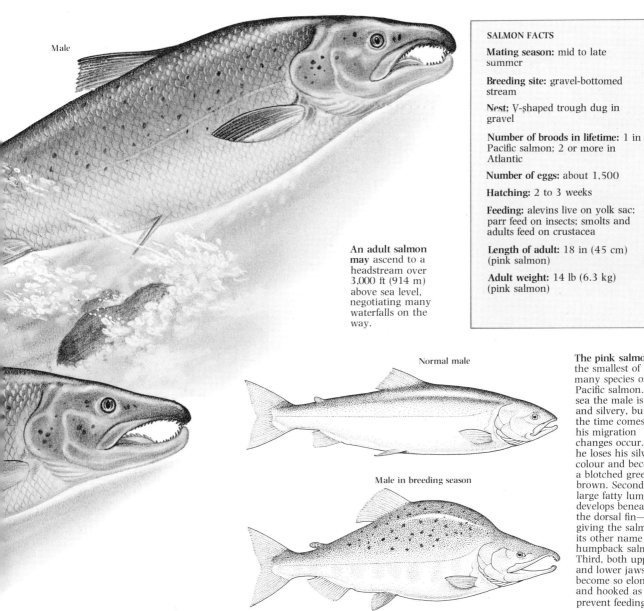

Male

An adult salmon may ascend to a headstream over 3,000 ft (914 m) above sea level, negotiating many waterfalls on the way.

Normal male

Male in breeding season

The pink salmon is the smallest of the many species of Pacific salmon. At sea the male is sleek and silvery, but as the time comes for his migration changes occur. First, he loses his silvery colour and becomes a blotched greeny brown. Second, a large fatty lump develops beneath the dorsal fin— giving the salmon its other name of humpback salmon. Third, both upper and lower jaws become so elongated and hooked as to prevent feeding.

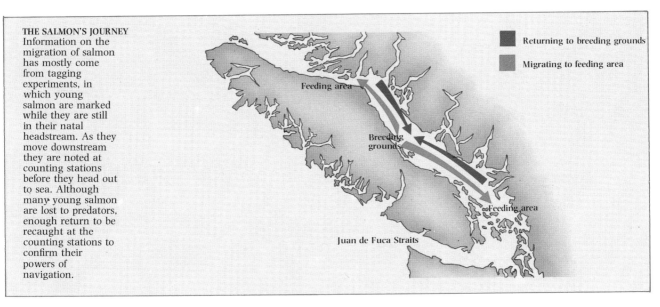

THE SALMON'S JOURNEY
Information on the migration of salmon has mostly come from tagging experiments, in which young salmon are marked while they are still in their natal headstream. As they move downstream they are noted at counting stations before they head out to sea. Although many young salmon are lost to predators, enough return to be recaught at the counting stations to confirm their powers of navigation.

Returning to breeding grounds

Migrating to feeding area

Feeding area

Breeding grounds

Feeding area

Juan de Fuca Straits

The crocodile

Crocodiles do not start to breed until they are about 17 years old, by which time they are 8 to 9 feet (2.4 to 2.7 metres) long. Bulls defend territories of up to 750 feet (230 metres) along a river bank or lake shore, and strangers are seen off with much jaw-snapping and roaring. The full-blooded roar sounds as if it comes from the base of a well and may last six or seven seconds. To the ears of the cows this aria is irresistible and several gather in the singing male's territory. He makes his intentions even clearer by raising his tail from the water, which submerges his body, and by blowing bubbles. His neck swells up and hissing air is expelled from his mouth and nostrils. As his excitement grows his whole body trembles and churns the water into a creamy froth. Much jaw-snapping and snorting ensues and the noise is increased as the male rears up out of the water and smashes his jaws down on to the surface. The tail is flailed from side to side, whipping the already turbulent water into a fury.

Impressed by this show of desire, the female, who has watched the spectacle passively, enacts a brief nuptial display and with no further ado mates with the male in the water. Mating signals the start of several months of devoted maternal behaviour—the male plays no further part in rearing his offspring. First, the female seeks out a nest site—an area of soft soil near shade where she can keep a watchful eye. Using her front limbs she digs a nest and lays 48 to 60 eggs. After covering the nest with soil she retires to the comfort of the shade. For three months she stands guard, never feeding or going far from the nest, but despite this, monitor lizards, hyenas, mongooses and vultures rob momentarily unguarded nests.

When incubation is complete, the baby crocodiles call to their mother from inside their egg shells. She excavates the nest, cracking and tearing apart the hard-baked covering soil. If, for any reason, she does not open the nest the young cannot escape, and they die in the chamber. As the hatchlings, which are about a foot (30 centimetres) long, struggle to leave their shells, the cow gently picks them up in her huge jaws and carries them to a quiet, sheltered nursery site. Here she continues her guard against predators, such as marabou storks and great white egrets, and for another six months devotes herself to the welfare of her young. By the time they are some 18 inches (45 centimetres) long and feeding on insects and small fish, the cow leaves them to turn her head once more in the direction of a thundering roar.

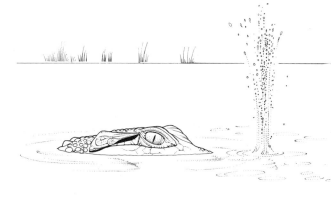

At the peak of his courtship excitement the bull crocodile makes a 'fountain display'. Filling his lungs with air, he plunges his nose under water and snorts through his nostrils. So great is the pressure of the air that a column of water is shot several feet up into the air. The 'fountain' plays for 5 or 6 seconds and the display may be repeated several times.

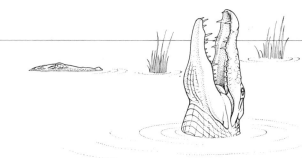

The female's role in courtship is mostly passive but she does have one ecstatic burst of activity. She performs her nuptial dance in the shallows, making rearing movements in which her gaping jaws point skywards. To the accompaniment of a wheezing, throaty groan, she snaps her jaws and rears up again. She displays for some minutes.

Once the mother crocodile has broken into the nest and released her babies, she shows her most remarkable maternal behaviour. The hatchlings are extremely vulnerable and the female cannot attend to all 40 or 50 of them at once. As they emerge from the eggs the young make for the nearest cover, but many are snapped up by waiting marabou and hornbills. Others shelter under the mother, and these

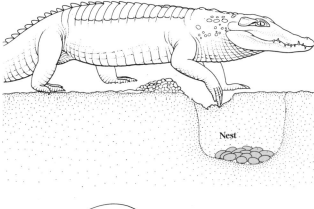

Nest

Mothers lay their **eggs** in nests near standing water in soil or sand. Nests are between 8 and 18 in (20 to 45 cm) deep. When she has replaced soil over the eggs the female crocodile lies directly on top of the nest area. During the 3 months of incubation she seldom strays from the nest. Even the greatest provocation does not displace her.

Inside their shells young crocodiles become sensitive to noise and vibration. When they are ready to hatch they respond to the foot-falls of their mother by calling out. They call until the cow breaks the hard-packed soil over the nest. Using its egg tooth each young chips its way out of the egg. The mother may help by pulling the babies free.

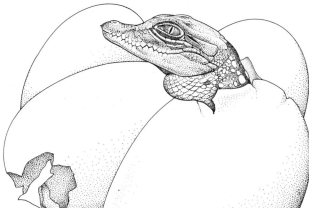

CROCODILE FACTS

Courtship: lasts 1 or 2 weeks

Mating season: July to September in Central and East Africa

Number of eggs laid: 48 to 60

Incubation: 12 to 13 weeks

Size when hatched: about 12 in (30 cm)

Period in nursery area: up to 6 months

Rate of growth: 10 in (25 cm) a year up to 6 years, then slows down

Feeding: hatchlings feed on insects

Sexual maturity: males do not breed until 9 ft (2.7 m) long and about 17 years old, and females breed when about 8 ft (2.4 m) long

Babies

she picks up first and transports the short distance to a safe stretch of quiet water. She makes several journeys from the nest site to the nursery site, each time with wriggling hatchlings in her jaws. She does not use her forelimbs to lift them but rather snaps the young up as if she were eating tasty prey. She does not close her jaws—the tiny crocodiles look like prisoners behind the bars of her teeth.

In the nursery the young crocodiles bask on their mother's head or tail and seem disinclined to enter the water for some weeks. The female is fierce and quite athletic in her defence of her babies.

143

The herring gull

Every year around March and April the herring gull, *Larus argentatus*, becomes interested in occupying a breeding territory on cliffs or sand dunes close to the sea. The males will defend a small area from rivals by driving them away and by fighting if necessary. But fighting can be a dangerous pastime— even a victor may risk serious injury which could lead to the loss of a territory, a mate, or even of life.

Herring gulls, like many other animals, have evolved a complex system of signals as a substitute for real fighting. Most common of these is the upright threat display. The neck is stretched up as if to peck, and the wings held ready as if to beat an opponent. Males will strut and posture for hours in this way. If the message is still not clear, males may peck and pull violently at the grass in a redirected attack.

Once a territory has been established, females will appear and tour the area looking for eligible males. Using an oblique posture and emitting long calls or shorter 'mews', the males call to the females. During courtship the female is extremely wary, and to convince her prospective mate that she comes in peace, often adopts a posture in which her chief weapon, the beak, is pointed away from him. If a male decides to accept her he may feed her and also indulge in a 'choking' display which is performed by both birds.

Both birds help with building a rough, twig-lined nest cavity in the sand. In it, the female lays about three eggs and both parents help in the task of incubation. After some 25 days the eggs hatch and the young are soon looking for their first meal. To get it they must peck at the red spot on their parent's lower beak. This stimulates the adult to regurgitate a meal of half-digested fish, which is quickly devoured by the young.

The young gulls reach full size in their first year, moulting their chocolate-coloured down for their first juvenile plumage. Now they learn to fly. They flap their wings frantically to begin with to develop muscular strength, but it is only by taking off on a real flight that they can truly practise. Initially there are many falls to the ground, clumsy landings and other disasters, but eventually the young can join their parents on the feeding grounds. There is still much to learn—how to catch fish and other food, where to sleep at night and how to avoid predators.

The next spring the young return to the breeding areas, and although they may not mate for several years will watch and wait until a suitable territory becomes available, then embark on the next stage of their life cycle.

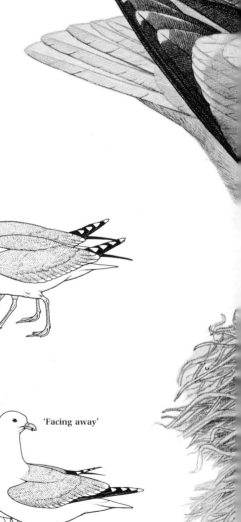

Before achieving full adult plumage, herring gulls undergo several moults. A juvenile in its first winter has brown plumage with darker primary feathers and tail, and a black bill. A second year bird has a grey back, and a tail with a white base and dark tip. The mature adult has a grey back and grey wings with black tips. The juvenile birds' plumages are a signal to adult birds that they are not yet mature and thus no competition.

In the courtship or meeting ceremony of herring gulls the male first calls to the female while in the oblique posture with head and bill upward. Both male and female may then give the 'mewing' call together to strengthen their bond. The 'mew' call is given with the head stretched forward and downward and the bill open wide.

Oblique posture

'Mew' call

The 'choking call' is also often made by both herring gulls, particularly when they are deciding on a nest site. Each gull makes a strange, rhythmical sound while moving its breast in time to the call. The 'facing away' posture reassures each gull of its partner's peaceful intentions —each points its main weapon, the bill, away from the other bird.

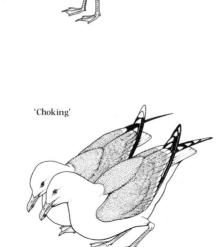

'Choking'

'Facing away'

First winter plumage

Second winter plumage

Adult plumage

HERRING GULL FACTS

Number of eggs: average clutch 3 eggs

Size of egg: 2.75 in (7 cm) by just under 2 in (5 cm)

Incubation period: 25 to 33 days

State at birth: born fairly well developed but fed by parents

Flight: able to fly at 6 weeks

First breeding: 3 to 5 years

Food: earthworms, shrimps, molluscs, starfish, fish, crabs

Length of adult: 22 to 26 in (56 to 66 cm)

The upright threat posture is the signal most commonly used by a herring gull to warn a rival male off his territory. He stands tense and rigid, his neck stretched and wings slightly raised as if ready for action. Holding himself in this posture he moves toward the intruder. Although his appearance may not seem dramatic to humans, it conveys a strong and unmistakable message to the other gull.

Red spot

A parent herring gull does not feed its hungry young spontaneously. To obtain food the young must peck at the red spot on the tip of the parent's lower bill. This stimulates the parent to regurgitate half-digested food easily eaten by the young.

The grey wolf

For both man and beast, life in the high Arctic is difficult indeed, but the wolf has adapted to it well. The reasons for the wolf's success lie in its strong family bonds and unified family action.

The leader of the wolf family is an adult male. His main responsibility is to provide and maintain a territory, which may be between 10 and 100 square miles (25 and 250 square kilometres). Together with his mate, to whom he remains paired for life, he patrols the territory boundary, urinating on prominent objects and at path intersections. The odour deters strangers.

Wolves pair when they become sexually mature—males at three and females at two years of age. In mid-February the female comes on heat for just two or three days. Special odorous substances in her urine tell her mate of her condition, and a brief courtship ensues. The dog approaches the bitch from the rear and sniffs her urinogenital region. Before the peak period in her receptivity has arrived the bitch responds to this attention by moving away, but when quite ready she adopts the mating posture with her hind legs slightly apart and her tail pulled to one side. The dog mounts her from behind, his forepaws gripping her flanks. As in all dogs, copulation lasts for up to half an hour, for the tip of the male's penis expands within the bitch's vagina making withdrawal impossible. Locked together they sink to the ground and lie still until the penis has subsided. At this time the young, unmated wolves stand guard, for the coupled pair are highly vulnerable.

After a gestation of about 60 days a litter of three to eight cubs is born deep within a den chosen for its good cover and inaccessibility to possible predators such as eagles and other wolves. Blind and hairless, the cubs grow quickly on the mother's milk. By the twelfth day their eyes are open and at three weeks they venture into the world for the first time. Up to this moment their father, aunts, uncles and older brothers and sisters have had nothing to do with them, for the bitch actively repels any adults straying into the breeding chamber. But now the whole family assists in playing with the cubs, for play strengthens family relationships and ensures the trust necessary for cooperative hunting

When the cubs are three months old the family abandons the den site and moves to the summer hunting site. Caches of meat, carefully buried the year before, provide nourishment until the new hunting season starts. The cubs join the hunting parties and by the start of the next breeding season are fully independent.

Peace within the pack is maintained by a rigorously applied hierarchy. The pack leader is always a dog, but his mate also gains a high degree of respect, and both keep their tails high as a badge of rank. The number of subordinates depends on the amount of food within the territory. These wolves must not carry their tails as high as the leader. If food resources fail, the least dominant adults may be driven out and only allowed to feed on the stripped carcasses of the pack's kills. The tails of outcasts are never raised. Cubs and juveniles can do whatever they like and their tails may be up or down. Only when a cub matures does the tail position have any social meaning.

Dominant pair

Subordinates

Outcasts

Cubs and juveniles

Grey wolves, *Canis lupus*, base their success on the tight structure of the family. With the dominant pack leader all adults take part in play with the cubs. Active and athletic play teaches the cubs to assess distance, attack times and how to ambush—all skills upon which the future survival of the pack depends.

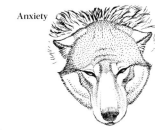

Anxiety

Suspicion

If strangers meet, both assume a subordinate posture and examine each other's urinogenital region. They urinate and sniff the urine. If one dog is dominant it lifts its tail while the other holds his between his legs.

Facial expressions are almost as valuable to wolves as they are to humans. Anxiety is expressed by a narrowing of the eyes and a backward movement of the ears. A suspicious wolf flattens its ears and makes its eyes mere slits.

WOLF FACTS

Courtship: wolves pair for life and courtship is brief

Mating season: mid-February. Female in heat 3 to 5 days

Gestation period: 60 to 62 days

Litter size: 3 to 8 pups

Weight at birth: about 1 lb (0.45 kg)

Eyes open: 12 days

Emerge from den: 21 days

Feeding: milk for 3 weeks; weaned on to insects, frogs and lizards

Weaning complete: 8 weeks

Sexual maturity: males at 3 years females at about 2 years

Adult weight of male: 95 to 100 lb (43 to 45 kg)

Adult weight of female: 80 to 85 lb (36 to 38.5 kg)

Shoulder height: 26 to 32 in (66 to 81 cm)

Colour: from white through cream, tawny, reddish-brown, grey and black according to sub-species

Play is important for the development of the cubs' social behaviour. As soon as they are able to romp about outside the den they start to discover their relative dominance positions. Play is a way of developing and expressing stamina and courage—the earliest traits of dominance. To these are added the degree of intelligence, boldness, timidity or viciousness shown by the cub, and the cub's size. Before the pack leaves the den site for the summer hunting grounds, the hierarchy in the litter begins to emerge. Barring accidents, this order is likely to persist into adulthood.

Cub in submissive posture

The baboon

Baboons are ground-dwelling monkeys living in Africa and extreme western areas of Asia. They have become adapted to life in open country, savanna and sparsely wooded grasslands, but may also be found in rain forests. Their diet is varied, consisting largely of fruits, grasses and roots, lizards and insects, and even the flesh of other mammals.

There are five species of baboon in the genus *Papio*. The gelada, an animal similar in appearance and behaviour to the 'true' baboons, is placed in a different genus, *Theropithecus*.

The life of a baboon centres around its position in a highly organized society. Baboons generally live in family groups of some 40 individuals (groups as large as 200 have been recorded) whose activities and movements are directed by a few fully grown dominant males. There is little mobility from one group to another. Dominance is established by aggressive behaviour—threats and attacks. Physical assault may involve biting at the nape of the neck and rubbing the victim in the dust, but although baboons are apparently violent, serious injury is rare. A dominant male usually need only resort to gestures in order to settle a dispute—he raises his eyebrows, blinks rapidly to display his light-coloured eyelids and bares his teeth. High rank may be gained not just by fighting ability, but

also by social skill in being able to form a cooperative association with another large male.

Females also form a hierarchy and are aggressive to each other and to juvenile or weak males. The social position of a female is difficult to determine; it is probably less stable than that of a male, as females are affected by physical and behavioural changes associated with their menstrual cycle. The baboon female shows marked changes in sexual receptivity during her 35 day cycle. When receptive she is usually closely attended by a dominant male, and this enhances her social status.

A day in the life of a baboon, however, is not spent constantly defending status. Minor threat postures are often sufficient to settle disputes, and dominant males keep a wary eye on the troop, ending a social disturbance with a glance. Most of the troop's time is spent peacefully feeding, or engaged in grooming and social behaviour.

But the hierarchies exist and are vital for the cohesion of the group. Each animal knows its place and exactly how to behave in relation to mating and looking for feeding or sleeping grounds. Baboon troops travel perhaps three or more miles a day in search of food, and there is little room for individualism. All activities must be directed and controlled by the older dominant males.

A dramatic threat posture is adopted by a male gelada in an aggressive encounter. The white upper eyelids are raised, the ears flattened and a bare area of red skin on the chest is expanded into full view. The baboon turns his lips back to expose the large, pointed canine teeth. This 'snarl' expression is all the more fearsome, as the upper lip is rolled back over the nose and the lower lip over the chin. When danger threatens, geladas often seek refuge in steep craggy terrain and hurl rocks and stones at their enemies.

The gelada baboon, *Theropithecus gelada*, of Ethiopia is a ground-dwelling monkey. It rarely, if ever, climbs trees, and sleeps on rock ledges, in caves or rock crevices. The male has a heavy, brown mane, covering his forequarters and is markedly larger than the female.

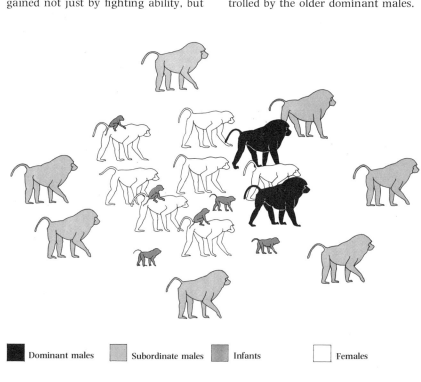

■ Dominant males   ▨ Subordinate males   ▨ Infants   ☐ Females

Social order determines the positioning of a baboon troop on the move. In the middle are the dominant males and mothers with infants. Subordinate males walk at the sides and rear, and the troop is ostensibly led by young adult males. Sub-adult males

scout the terrain as much as a quarter of a mile ahead of the main group. They do not determine direction and have to move fast if the dominants decide on a change. Danger is signalled by an alarm bark from a peripheral male.

BABOON FACTS

Gestation period: 6 months

State at birth: central nervous system comparable to 6 month old human infant

Feeding: mother's milk for about 8 months and solid food from 3 to 4 weeks

Tactile exploration: 3 days

Coordinated reaching: 3 days

Eye-hand-mouth coordination: 8 days

Movement: able to walk steadily at 3 weeks

Sexual maturity: female $3\frac{1}{2}$ to 4 years, male 4 to 6 years

Full development of dominant adult male: 10 to 11 years

Adult weight of male: 48 lb to 66 lb (22 kg to 30 kg)

Adult weight of female: 24 lb to 33 lb (11 kg to 15 kg)

Silent bared-teeth face

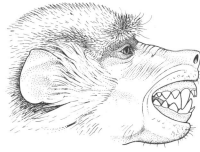

Staring open-mouth face

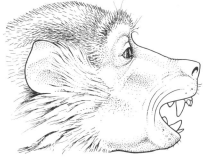

Facial expressions are a vital means of communication for baboons. The silent, bared-teeth face is shown by a subordinate on the approach of a dominant, perhaps to appease him. The staring, open-mouth face, accompanied by a staccato bark, is a threat expression given by a dominant.

The baboon/2

When the sexual skin on her rear shows signs of swelling and develops bright red colouring, a female baboon starts to approach males. At this stage of her menstrual cycle, however, she is ignored by all but juveniles and sub-adult males. Only during the week or so prior to ovulation, when her sexual swelling is at its greatest, is she attractive to dominant males. She forms a special bond with one dominant male and he repeatedly mates with her.

Pregnancy lasts about six months and during this period the female spends little time in the company of adult males. The birth itself may take seven hours; as soon as the baby is born the mother licks it clean and eats the placenta. The infant begins to feed at its mother's breast within 20 minutes of birth, and during its first few days of life rarely takes its mouth from her nipple.

When she has a baby a female baboon's social position changes radically. She moves to the heart of the troop close to the oldest, most dominant males. Here she enjoys their protection and is virtually immune from attack by other troop members. She can give her undivided attention to her infant, constantly licking, grooming and nuzzling it, and exploring its body.

The infant spends nearly all of its first month in its mother's arms. Toward the end of this time it is able to walk a little, and begins to respond to the constant attentions of other troop members. Nursing mothers tend to sit together during rest periods, and young infants tentatively begin to approach and briefly touch each other, always running back to mother when feeling insecure. The infant rarely leaves its mother for more than a few minutes until it is well into its third month. At about five weeks the infant first climbs on to its mother's back, and from two months it rides on her back, jockey style.

At three months the infant begins to spend an increasing amount of time playing with other infants of the same age. Older infants and juveniles are greatly interested in their juniors, and constantly entice them to play rough-and-tumble games. Adult males keep a close watch on their play and quickly intervene if any young show signs of distress.

Between four and six months the fur on the infant's belly and the sides of its face changes from black to brown. This marks the end of the first stage in the baboon's development. The infant now wanders far from the mother during a quiet feeding period but still rushes back to her at any sign of danger; it still rides on its mother's back to the sleeping trees and sleeps in her arms. It begins to take solid food and is probably weaned by eight to ten months.

The female baboon continues to care for and watch over her child until it is about two years old, when she may have another baby. This protection is important, as from eight months the juvenile may be roughly treated by other adult females. Dominant males of the central hierarchy, with whom the infant was closely associated in the early days of its life, continue to intercede on its behalf until it is about three years old.

A new-born infant arouses great interest in a baboon troop. Older juveniles and sub-adult females show the most concern and attempt to groom the mother and touch the baby. Young adult males show little interest, but older dominant males often confidently approach and touch the infant. Although a mother may cringe when the dominant male fondles her baby, he is gentle and careful.

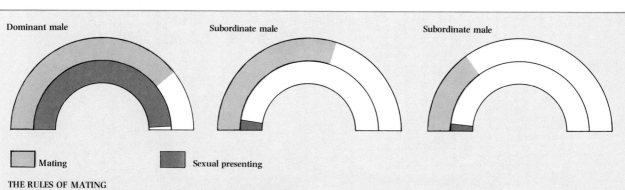

Dominant male Subordinate male Subordinate male

☐ Mating ▨ Sexual presenting

THE RULES OF MATING
Subordinate and sub-adult males may mate with a female baboon, but they are restricted to the early phase of her sexual receptivity long before she can actually conceive. Dominant males mate only with females at the height of the sexual swelling, when copulation is most likely to result in fertilization. The dominant male not only mates more often with receptive females than his subordinates do, but is also more often sexually presented to by females and juveniles. The genes of dominant males have therefore the best chance of being passed on to future generations of baboons.

Female

Fertile female

The skin on a baboon's rear is hairless and usually brightly coloured. In the female this region extends from the base of the tail to the junction of the abdominal wall with the front of the thighs. At the approach of the fertile phase of her menstrual cycle this area becomes grossly swollen and bright red. The sexual swelling lasts about 19 days and declines after ovulation. The female is particularly receptive during the last 7 to 10 days.

A dominant male mates with a female during her most fertile time, the last week or 10 days of her sexual swelling. They spend most of their time together during this period and he frequently mounts her, brooking no interference from subordinates. Copulation lasts only about 7 or 8 seconds, during which time the male makes up to 15 pelvic thrusts.

The lion

Unlike most cats, lions are not solitary but are social animals that live in groups called prides. The lion life cycle starts with the mating of male and female. When sexually mature, the female comes on heat (into estrus) every few weeks and in this condition is most attractive to males. One dominant male will mate with her—on average once every 15 minutes for several days. After a gestation period of some 14 weeks the female gives birth to a litter of two or three tiny cubs which suck milk from their mother for more than six months before being weaned on to raw meat.

For the next two or three years young males and females grow up in the pride, but then their paths diverge. Young females usually stay with the pride for life. Thus the permanent nucleus of the pride is a group of interrelated females —mothers, daughters, sisters, grandmothers, cousins and so on. A young female will start to produce cubs when she is about three or four years old.

Quite a different pattern of events lies in store for the young males. Once they are about three, the dominant males become increasingly aggressive toward them and may forcibly expel them from the pride, but sometimes two or three brothers just seem to feel restless and leave the pride of their own accord. Whatever the reason, all young males leave the pride of their birth and set off to seek their fortunes elsewhere. For several years they may wander through the bush, hunting for themselves and, as they do so, becoming bigger and stronger. They may mate with stray females, but are really on the lookout for a pride of their own. Young males may take over a pride in which the males have died, but if not will eventually challenge the males of an existing pride. If these males are getting old and weak, stronger males will win. Old toothless males cannot usually hunt for themselves, and starve to death.

Sometimes after a take-over the new males systematically kill off young cubs. This ruthless infanticide has a logical explanation. Male lions are not interested in genetically unrelated offspring. In a biological sense it is better to remove existing cubs and start the females producing their own cubs as soon as possible. There may soon be a day when the males are ousted from the pride, so there is no time to waste.

As long as the pride is rearing the young of its males and females, then the curious elaborate social contract between two unrelated groups of kin is maintained. But the pride is an uneasy alliance of females and males (often brothers) that will eventually be broken and a new one formed.

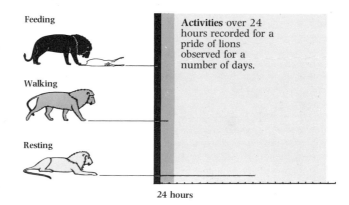

Feeding

Walking

Resting

Activities over 24 hours recorded for a pride of lions observed for a number of days.

24 hours

Walking, feeding and resting are the activities which make up the average day of a lion. Resting, which takes up most of the 24 hours, includes sedentary activities such as grooming and licking. Hunting activity is difficult to assess quantitatively. Only actual stalking or hunting movement can be measured accurately.

A lioness suckles cubs from any litter in the pride, although she gives her own young preference. This communal suckling means that cubs can feed even if their own mother is short of milk, and also that a female who has lost her cubs can use her milk for others.

A pride of lions, *Panthera leo,* may include 2 to 4 adult males, 3 to 12 adult females and a number of sub-adults and cubs. Pride numbers can vary enormously from 3 or 4 individuals to as many as 20 or 30. Adult females do most of the hunting, and look after and feed the cubs. A pride occupies a territory which may be several miles across.

The bared-teeth expression is a common signal among lions. A lioness may use a mild version to discourage a boisterous cub. She lays her ears back and retracts her lips slightly to show the teeth. Such an expression is really a warning to say that the exposed weapons might be used if necessary.

A lion is a cub until about 2 years old and a sub-adult until 4. A female in a pride usually has her first litter at the age of 4 and may continue reproducing for 12 years. Her life span, barring accidents, is about 18 years.

Cubs are dependent on adults for food until almost a year and a half old. After this they still need the help of adults to catch anything other than small prey.

Mother suckling another female's cub

Cubs have no contact with the pride for about 2 months and even after this age, play and exploratory activity usually take place in the reassuring presence of the mother. When she returns from a hunting trip, the cubs have a burst of activity, investigating their surroundings with a mixture of timidity and curiosity.

The animal kingdom

The relationships between different types of animals are shown in two separate family trees—non-chordates (animals without backbones) and chordates (animals with backbones). Both concentrate on animals described in this book. From top to bottom the categories change from broad groups including many different kinds of animal, to smaller, more specific groups.

PROTOZOA
Single-celled animals

PLATYHELMINTHES
Flatworms
Flukes
Tapeworms

MOLLUSCS
Chitons
Snails and slugs
Bivalve molluscs
Squid, cuttlefish
and octopus etc

ANNELIDS
Bristleworms
Earthworms
Leeches

CRUSTACEANS
Barnacles
Prawns, crabs
Copepods etc

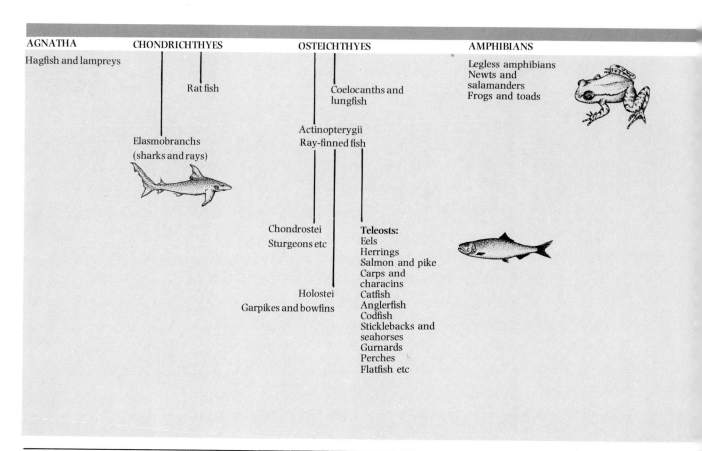

CHORDATE ANIMALS

NON-VERTEBRATES
Sea squirts
Lancelets etc

AGNATHA
Hagfish and lampreys

CHONDRICHTHYES
Rat fish

Elasmobranchs
(sharks and rays)

Chondrostei
Sturgeons etc

Holostei
Garpikes and bowfins

OSTEICHTHYES
Coelocanths and lungfish

Actinopterygii
Ray-finned fish

Teleosts:
Eels
Herrings
Salmon and pike
Carps and characins
Catfish
Anglerfish
Codfish
Sticklebacks and seahorses
Gurnards
Perches
Flatfish etc

AMPHIBIANS
Legless amphibians
Newts and salamanders
Frogs and toads

ARTHROPODS (13 CLASSES)

ECHINODERMS

Starfish
Brittle stars
Sea urchins
Sea cucumbers
Sea lilies

ARACHNIDS

Scorpions
Spiders
Daddy-long-legs
Whip scorpions etc

INSECTS

Mayflies
Dragonflies
Cockroaches
Mantids
Termites
Grasshoppers
Stick-insects
Bugs
Beetles
Fleas
Ants, bees and wasps
True flies
Butterflies and moths
etc

VERTEBRATES

REPTILES

Turtles and tortoises
Tuatara
Lizards and snakes
Crocodiles

BIRDS

Kiwis
Ostrich
Rheas
Cassowaries and emus
Tinamous
Grebes
Loons
Penguins
Albatrosses, shearwaters and petrels
Pelicans and allies
Herons, storks and flamingoes
Screamers and ducks
Eagles, hawks and vultures
Game birds
Cranes and allies
Waders, gulls and auks

Pigeons
Parrots
Cuckoos and turacos
Owls
Frogmouths and nightjars
Swifts and hummingbirds
Trogons
Colies
Kingfishers, hornbills and allies
Woodpeckers, barbets and toucans
Passerines

MAMMALS

Monotremes:
Echidna
Duck-billed platypus

Marsupials:
Kangaroos
Bandicoots
Wombats
Opossums etc

Placentals:
Insect-eaters
Colugos
Bats
Primates
Whales
Anteaters, sloths and armadillos
Pangolins
Pikas, hares and rabbits
Rodents
Flesh-eating mammals
Seals

Aardvark
Elephants
Hyraxes
Sea-cows
Odd-toed hoofed mammals
Even-toed hoofed mammals

Index

spotted, 54
Hyla citroba, 41
Hypolimnas misippus, 21
Hypothalamus, 26
Hypsignathus monstrosus, 70

I

Icteridae, 132
Impala, *113*
Imprinting, 117, 130, *131*
Incubation, 97, 99, *101*, 102–3, 108, 110, *110*, 111, 116, 130, 142
Indicatoridae, 132
Infundibulum, *29*, *100*
Insect, 10, 18, 22, 24, 40, 60, 74, 97
Insectivore, 62, 108
Invertebrate, 24, *24*, 25, 26, 32, 44, 60, 64, 74, 104, 130
Isoodon obesulus, 110
Isthmus, *29*, *100*

J

Jacobson's organ, 46, *47*
Jellyfish, 96

K

Kalotermitid, 78
Kangaroo, 50, *83*, 110, *110*, *111*
  red, *118*
Keratin, 100
Kidney, *29*
Killdeer, *125*
Kingfisher, *86*
Kiwi, 46, *101*, 102
Koala, 110
Kudu, *51*

L

Labour, 114
Labridae, 24
Labroides dimidiatus, 24
Laniarius aethiopicus, 58
Larus argentatus, 144
Larvae, 106, *106*, *107*
Learning, 117, *127*, 128, 130, *130*
Leipoa ocellata, 88
Lekking, 65, 70, *70*, 71, 72, *72*
Lemur, 58
Leopard, 50, *51*, 66
Lepidoptera, 44
Leporillus conditor, 90
Limax maximus, 24
Limpet, 106
  slipper, 24, *25*
Ling, 18, 28
Lion, 17, *21*, 30, 33, 66, 122, *129*, 134, 152–3
Lizard, 32, 36, *36*, 82, 92, 97
Lobster, 44
Luciferin, *34*
Luteinizing hormones (LH), 27, *27*
Lycaon pictus, 92, *122*

M

Macaca mulatta, 126
Macropodus spp., *80*
Macrotermes bellicosus, 78, 79
Magnum, *100*
Mallard, 38, *39*
Mammal, 26, 30, 33, 60, 112–3, 114, 117, 118, 130

arboreal, 94
bonding, 54
breeding seasons, 28
burrowing, 64, 75, 92
care of young, 104, 116, 118, 119, *122*
chromosome pattern, 22, *22*, 23, *23*
sex organs, 28, *29*, 62
see also Marsupial; Monotreme
Mammary ducts, *119*
Mammary glands, 27, 33, 110, *111*, 116, 118, *118*, *119*, 126
Man, 10, 14, 22, *22*, 26, 27, 28, 40, 54, 55, 56, 62, *62*, *63*, 74, 112, *113*, 135
Manakin, 65
Mandarin, *39*
Marine worm, 24
Marmoset, 82
Marsupial, 46, 62, 74, *82*, 94, 110–1, 122, 134
Masseter muscles, *50*
Mating, *19*, 24, 27, 42, 46, *46*, 60–2, 134
Mating calls, *19*, *41*
Mating postures, 46, *46*, 50, 146
Mayfly, 30, 106
Megaloceros, 20
Megapode, 88, 120
Megapodius freycinet, 88
Megaptera novaengliae, 40
Meiosis, 8, 9, 12, *12*, *13*, 14, 15
Meles meles, 128
Menstrual cycle, 27, 150
Metamorphosis, 106, *106*
Micromys minutus, 94, 94
Microstomum, 11
Milk, 116, 118, *118*
Milk ducts, *119*
Milk glands, 104, 118, *119*
Milk line, 118, *118*, *119*
Mimicry, 55, 132, *133*
Mite,
  itch, *31*
Mitosis, 10, 12, *12*, *13*
Mole, 75, 92
Mollusc, 18
  marine, 106
Mongoose,
  marsh, 68
Monkey, 32, 82, 94, 117
  howler, 64, *69*
  macaque, 58
  rhesus, 126, *126*, *127*, 135
  vervet, 58
Monogamy, 33, *48*, 52, *52*, 120
Monotreme, 62, 110–1
Moose, 122
Mosquito, 40, 60
Moth, 32, 40, 44
  pyralid, 44
Mother-infant relationship, 126
Mouse, 92, *113*
  deer, 52, *131*
  harvest, 75, 94, *94*
Mud puppy, 106
Muskrat, 90
Mussel, 106
Mustelus canis, 104

N

Nannopterum harrisi, 56
Natural selection, 16, 56, 66, 96
Nectophyrnoides, 28
Necturus, 106

Neoteny, 106
Nest,
  ants, 76, *76*, *77*
  bees, *136*
  birds, 74–5, 84, 85, 86, *86*, *87*, 88, *88*, *89*, 102, 108, 116, 144
  mammalian, 94, 116
  termites, 78, *78*, *79*
Neural groove, 98, *99*
Nipples, 110, *119*
Nocturnal animals, 32, 34, *34*, 40

O

Octopod, 60
Octopus, *19*, 26, 60, 97
Oecophylla, 76
Oncorhynchus gorbuscha, 140
Ondatra zibethica, 90
Oocyte, 15
Ophryotrocha puerilis, 24
Orang-utan, *22*, 63
Orgasm, 54
Ornithorhynchus anatinus, 111
Orpendola, 132
Orycteropus afer, 92
Oryx, 66, *67*
Oryx gazella, 67
Os penis, 18, 62, *62*
Ostrich, *101*, *102*
Otaria byronia, 53
Otter, 53, 68
Ovary,
  bird, 28, *100*
  fish, 29, 46, *47*
  frog, *29*
  mammal, 14, 28, 29, 112, *112*, *113*, 114
Ovenbird, 75, 86, *86*
Oviduct, 25, *29*, 62, 100, 104, *113*
Oviparity, 104
Ovotestis, *24*
Ovulation, 27, *151*
Ovum, 8, 14, 27, 98, 100, 112
  division of, 98
  fertilization of, 28, *29*, 112
Owl, 65, *102*, 120
  burrowing, *85*, *93*
Oxytocin, 114, 118

P

Pairing *see* Bonding
Panda, 109
  giant, 109
Pangolin, 82
Panthera leo, 152
Panthera pardus, 51
Panthera tigris, 19
Pan troglodytes, 22, 57
Papilla, 62
Papio, 148
Paradisaea raggiana, 35
Paramecium, 8, 9
Parapodia, 60
Parasitism, *10*, 20, 24
  see also Brood parasitism
Parental care, 28, 52, 54, 55, 56, 80, *80*, 88, 104–5, 118
Parent-young recognition, *130*
Parnassius, 18
Parthenogenesis, 9, 10
Partridge, 108
Parus ater, 131
Parus caeruleus, 131
Passerine, *102*, 108, 120

Acknowledgments

The authors contributed text as follows:
Dr Clive K. Catchpole: 16–7, 32–43, 74–9,
84–9, 116–7, 120–33, 136–9, 144–5, 152–3

Dr Frank H. Hucklebridge: 8–15, 18–31,
96–109, 112–5, 118–9, 148–51

Dr D. Michael Stoddart: 44–73, 80–3, 90–5,
110–1, 134–5, 140–3, 146–7

The Publishers received invaluable help
during the preparation of *The Animal
Family* from:
Ruth Binney, Marsha Lloyd and John
Porter, who gave editorial assistance; Mark
Collins of the Council of Overseas Pest
Research; Ann Kramer, who compiled the
index.

The following books and articles were
particularly useful during the preparation
of *The Animal Family*:
The Animal and its World (volume 1), N.
Tinbergen, George Allen & Unwin, London
1972
Animal Architecture, Karl von Frisch,
Hutchinson, London 1975
Badgers, E. Neal, Blandford Press, Dorset
1977
Birds of Paradise and Bower Birds, Thomas
Gilliard, Weidenfeld and Nicolson, London
1969
The Cheetah, R.L. Eaton, New York 1974
The Cichlid fishes of the Great Lakes of Africa,
G. Fryer and T.D. Iles, Oliver and Boyd,
Edinburgh 1972

The Dancing Bees, Karl von Frisch,
Methuen, London 1966
Fish Migration, F.R. Harden Jones, Edward
Arnold, London 1968
How Animals Communicate, Thomas A.
Sebeok, Indiana University Press,
Bloomington and London 1977
The Insect Societies, E.O. Wilson, Cambridge,
Mass. 1972
An Introduction to Embryology, B.L.
Balinsky, W.B. Saunders, Philadelphia
1975
The Life of Birds, J.C. Welty, Constable,
London 1962
The Life of Mammals, J.Z. Young, Oxford
1975
Mammalogy, H.L. Gunderson, McGraw-Hill,
USA 1976
Maternal Behaviour in Mammals, H.L.
Rheingold, John Wiley & Sons, London
1963
A New Dictionary of Birds, Sir A.
Landsborough Thomson, Nelson, London
1964
Sociobiology, E.O. Wilson, Harvard
University Press 1975
The Serengeti Lion, G.B. Schaller, University
of Chicago Press 1972
The Wolf, L.D. Mech, New York 1970

Typesetting by Servis Filmsetting Ltd.,
Manchester
Origination by Acolortone Ltd., Ipswich
Printed by Printer Industria Gráfica S.A.,
Barcelona